Applied Mathematical Sciences | Volume 7

Applied Mathematical Sciences

F. John
PARTIAL DIFFERENTIAL EQUATIONS
ISBN 0-387-90021-7

L. Sirovich
TECHNIQUES OF ASYMPTOTIC ANALYSIS
ISBN 0-387-90022-5

J. Hale
FUNCTIONAL DIFFERENTIAL EQUATIONS
ISBN 0-387-20023-3

J. K. Percus
COMBINATORIAL METHODS
ISBN 0-387-20027-6

K. O. Friedrichs and R. von Mises
FLUID DYNAMICS
ISBN 0-387-20028-4

W. Freiberger and U. Grenander
A SHORT COURSE IN COMPUTATIONAL PROBABILITY AND STATISTICS
ISBN 0-387-20029-2

A. C. Pipkin
LECTURES ON VISCOELASTICITY THEORY
ISBN 0-387-20030-6

A. C. Pipkin

Lectures on Viscoelasticity Theory

With 16 Illustrations

Springer-Verlag New York · Heidelberg · Berlin

1972

Allen C. Pipkin

Division of Applied Mathematics
Brown University
Providence, Rhode Island

Library of Congress Catalog Card Number 74-186996.
Printed in the United States of America.

ISBN 0-387-90030-6 Springer-Verlag New York • Heidelberg • Berlin
ISBN 3-540-90030-6 Springer-Verlag Berlin • Heidelberg • New York

TABLE OF CONTENTS

Lectures on Viscoelasticity Theory

INTRODUCTION

This book contains notes for a one-semester course on viscoelasticity given in the Division of Applied Mathematics at Brown University. The course serves as an introduction to viscoelasticity and as a workout in the use of various standard mathematical methods.

The reader will soon find that he needs to do some work on the side to fill in details that are omitted from the text. These are notes, not a completely detailed explanation. Furthermore, much of the content of the course is in the problems assigned for solution by the student. The reader who does not at least try to solve a good many of the problems is likely to miss most of the point.

Much that is known about viscoelasticity is not discussed in these notes, and references to original sources are usually not given, so it will be difficult or impossible to use this book as a reference for looking things up. Readers wanting something more like a treatise should see Ferry's *Viscoelastic Properties of Polymers*, Lodge's *Elastic Liquids*, the volumes edited by Eirich on *Rheology*, or any issue of the *Transactions of the Society of Rheology*. These works emphasize physical aspects of the subject. On the mathematical side, Gurtin and Sternberg's long paper *On the Linear Theory of Viscoelasticity* (ARMA **11**, 291(1962)) remains the best reference for proofs of theorems.

Here we tend to emphasize mathematics for its own sake, but in the sense of a language for everyday use rather than as a body of theorems and proofs. No theorems are proved. Many problems are formulated, and some are more or less solved.

In Chapter I the basic material properties defining viscoelastic response in shear are defined and discussed. These properties appear in the form of response functions rather than as coefficients or parameters. Shearing response can be specified by giving either the stress relaxation modulus, the creep compliance, the complex modulus, or the complex compliance. The first problem in viscoelasticity theory is to find the relations specifying any one of these functions in terms of any other. In Chapter I we give the exact relations, and in Chapter III we consider approximate relations, which are more useful in practice. Since the relations are most easily stated in terms of Fourier and Laplace transforms, in Chapter II we briefly outline these transform methods.

Throughout these notes, basic physical laws are introduced only when they are needed. Since they are not needed in the first three chapters, they are not mentioned. The momentum equation appears first in Chapter IV, in a one-dimensional form suitable for the problem under discussion. Since we encounter non-trivial problems even in connection with quasi-static oscillations and one-dimensional wave propagation, we have not thought it necessary to obscure the view by introducing masses of equations that cannot be used.

In Chapter V the fully three-dimensional form of the momentum equation is introduced, but it is immediately specialized to the case of quasi-static motions, in which it is an equilibrium equation. At the same time, three-dimensional forms of the constitutive equations are introduced for the first time. In this chapter, which is on stress analysis, we can afford to write down general equations because we do not intend to solve them; the object of this chapter is to show how stress analysis problems in viscoelasticity theory can be reduced to problems in elasticity theory. Techniques for solving elasticity problems are not discussed. In these notes, we emphasize what is peculiar to viscoelasticity and leave aside topics that can be treated more thoroughly in courses on classical elasticity or fluid dynamics.

It is in keeping with this philosophy that in Chapter VI, on thermal effects, most of the discussion concerns the effect of temperature on relaxation times, a topic that doesn't arise in elasticity theory or Navier-Stokes fluid dynamics. Here the energy equation is introduced for the first time, but it is immediately discarded in favor of a specialized and approximate version that can be handled in the problem under discussion.

The first six chapters give an outline of linear viscoelasticity theory. In the remaining three chapters we consider some problems involving large deformations, particularly flow problems. The first difficulty in such problems is in the choice of a constitutive equation. Chapter VII concerns cases in which the material response is linear but the kinematical description is not, because strains are small but displacements are large. Chapter VIII concerns problems that can be linearized because the flow is close to Newtonian. Chapter IX concerns viscometric flows, in which the response is fully non-linear, but tractible because there are no transient time effects. No general theory of non-linear response is discussed because the general theory is too vague to be of much help in solving problems.

CHAPTER I

VISCOELASTIC RESPONSE IN SHEAR

In the classical linearized theory of elasticity, the stress in a sheared body is taken to be proportional to the amount of shear. The Navier-Stokes theory of viscosity takes the shearing stress to be proportional to the rate of shear. In most materials, under appropriate circumstances effects of both elasticity and viscosity are noticeable. If these effects are not further complicated by behavior that is unlike either elasticity or viscosity, we call the material viscoelastic.

From the broader and more unified point of view that the theory of viscoelasticity affords, we will be able to see that perfectly elastic deformation and perfectly viscous flow are idealizations that are approximately realized in some limiting conditions. For some materials it is these limiting conditions that are most easily observed. The elasticity of water and the viscosity of ice may pass unnoticed. In describing the behavior of materials mathematically, we use idealizations that depend strongly on the circumstances to be described, and not only on the nature of the material. We will find that the distinctions between "solid" and "fluid" and between "elastic" and "viscous" are not absolute distinctions between types of materials.

1. Stress Relaxation.

Now, and for some time to come, we consider the behavior of a slab of material in simple shearing motion. The slab is to be regarded as so thin that

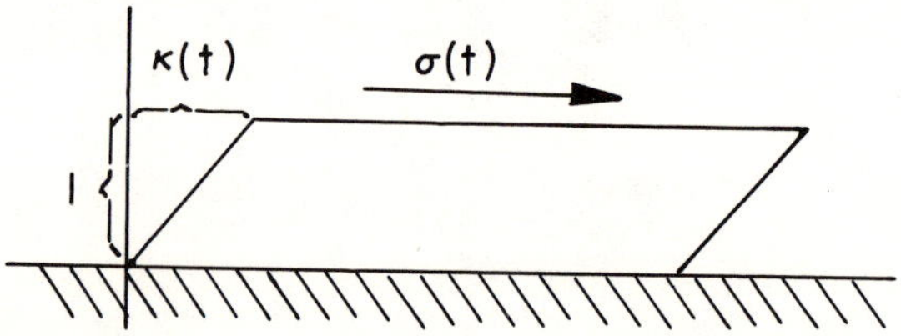

inertial effects can be ignored. Then the slab can be regarded as homogeneously deformed, with the amount of shear $\kappa(t)$ variable in time. Let $\sigma(t)$ be the shearing stress, the force per unit area on the slab.

We first consider the single-step shearing history $\kappa(t) = \kappa_0 H(t)$. Here $H(t)$ is the Heaviside unit step function, zero for negative t and unity for t zero or positive. If the material were perfectly elastic, the corresponding stress history would be of the form $\sigma(t) = \sigma_0 H(t)$, constant for t positive. If the material were an ideal viscous fluid, the stress would be instantaneously infinite during the step, and then zero for all positive t, like a Dirac delta, $\delta(t) = H'(t)$.

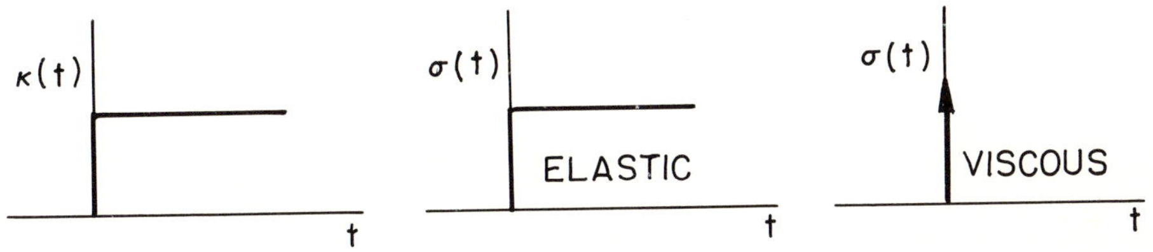

Close observation of real materials shows that neither of these idealizations is quite accurate. The stress usually decreases from its initial value quite rapidly at first, and later more gradually, approaching some limiting value $\sigma(\infty)$.

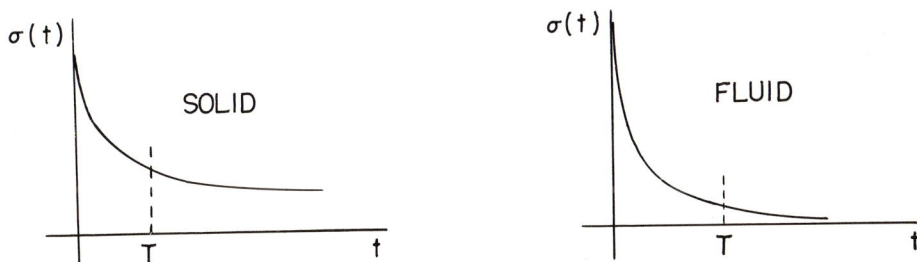

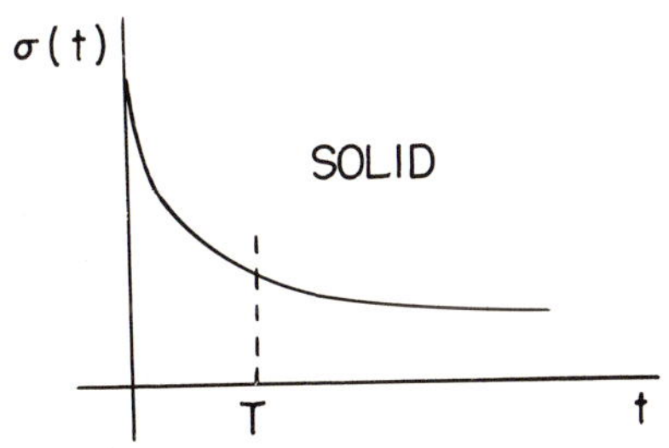

The limiting value is a subjective matter, since no one can wait that long, but it is a convenient idea. If the limiting value is not zero, we are likely to call the material a solid, and if it is zero, and the approach to zero is sufficiently rapid, we call the material a fluid. Evidently the nomenclature depends on an interaction between the nature of the material and the nature of the observation.

Let T be the relaxation time. Here, we provisionally take T to be some order-of-magnitude estimate of the time required for stress relaxation to approach completion. The distinction between solid and fluid is usually made on the basis of a subjective comparison of the relaxation time and the time of observation. If you can dabble your fingers in a material, you will probably call it a fluid even though it may return exactly to its initial configuration after a week or a month. If you are hit on the head by a hard object, you will think that it is a solid even if it can flow on a geological time scale.

Silly putty seems so strange because its relaxation time is so commensurable with our attention span. It will bounce, like an "elastic solid", the process being complete before there is time for much stress relaxation. It will also flow, like a "viscous fluid", before human patience runs out.

In the stress relaxation experiment on a sheared slab, the relaxation time T may be so short that it escapes observation, and the experimenter may then conclude that he is dealing with a perfectly elastic solid or a fluid, as the case may be. If T is so long that no stress relaxation is observed during the period of the experiment, again the observer may conclude that the material is perfectly elastic. We bother to call materials viscoelastic, and use appropriate mathematical models, when the relaxation time and the period of observation are not astronomically different.

2. Creep.

Now suppose that a slab is subjected to a one-step stress history $\sigma(t) = \sigma_0 H(t)$. The response of an elastic solid would be $\kappa(t) = \kappa_0 H(t)$, constant shear for t positive. In a viscous fluid, the shear would increase at a constant rate,

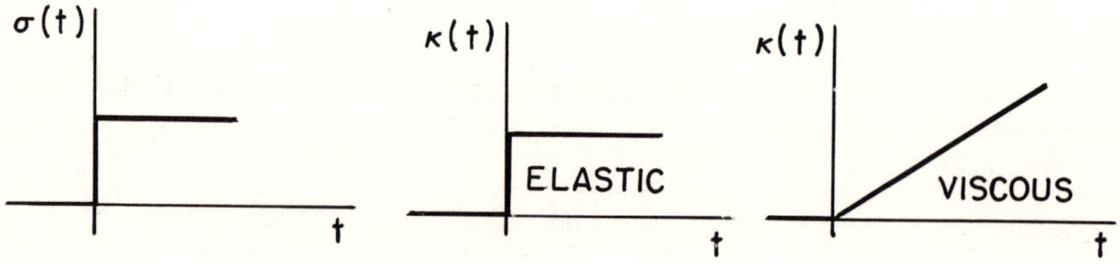

$\kappa(t) = \sigma_0 t/\eta$, the coefficient η being the viscosity. Viscoelasticity theory recognizes more refined observations which show departures from these idealizations. The shear at first jumps, so far as anyone can tell, so that the instantaneous response is elastic. The shear then continues to increase, but at lower and lower

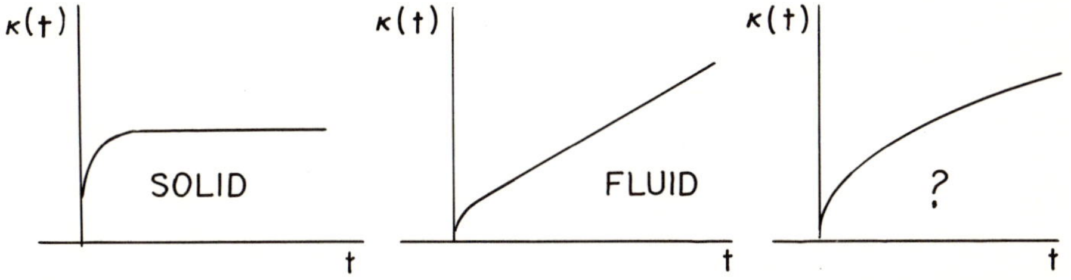

rates. If it appears to approach a limiting value $\kappa(\infty)$, the material is said to be solid. If it appears to increase linearly after a long time, the material is called fluid. It can easily be impossible to tell which type of behavior is occurring, if either. For example, if the amount of shear were increasing in proportion to $t^{\frac{1}{2}}$, one might become convinced by looking at limited data that a limiting value had been reached, or that the shear were increasing linearly, depending on one's preconceived ideas.

3. Response Functions.

Let $R(\kappa,t)$ be the stress relaxation function, the stress t units of time after application of a shear step of size κ. Let $C(\sigma,t)$ be the creep function, the shear t units of time after application of a stress σ. R and C are zero for t negative.

If the material is isotropic (has no distinguishable directions), it is evident by symmetry that R is an odd function of κ and C is odd in σ. Hence, assuming smooth dependence and supposing that κ and σ are small, we have

$$R(\kappa,t) = G(t)\kappa + O(\kappa^3)$$

and

$$C(\sigma,t) = J(t)\sigma + O(\sigma^3).$$

The coefficients of the linear terms are the linear stress relaxation modulus $G(t)$ and the linear creep compliance $J(t)$.

The values of these functions at $t = 0+$ are denoted G_g and J_g (g for glass), and the values at $t = \infty$ are G_e and J_e (e for equilibrium), provided that these values exist. If $J(t)$ tends to increase like $(t+T)/\eta_0$ for large t, η_0 is called the steady-shearing viscosity and T is the mean relaxation time.

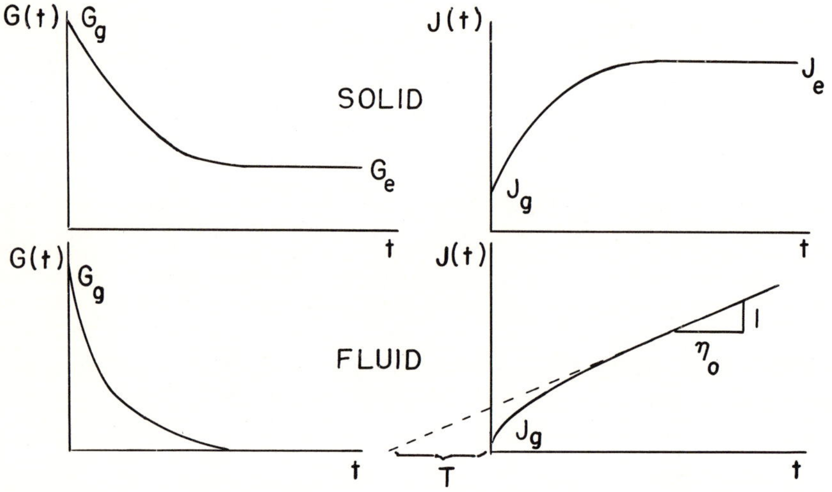

Immediately after application of a step in stress or strain, the response is independent of whether it is the stress or the strain which is to be held constant in the future. Hence, at time 0+, $\sigma = G_g\kappa$ and $\kappa = J_g\sigma$. Thus, we find that the initial values of G and J are reciprocal:

$$\boxed{J_g G_g = 1.}$$

If the stress and strain approach limiting values after a long time, for viscoelastic materials it is irrelevant which one was held absolutely constant for all positive time. Thus, at time infinity both $\sigma = G_e\kappa$ and $\kappa = J_e\sigma$ are valid.

Hence,

$$\boxed{J_e G_e = 1.}$$

We shall find that $J(t)$ and $G(t)$ are roughly reciprocal at all times, although exactly so only in the two limiting cases mentioned, and in the case of perfectly elastic response. The reciprocal relations for the two limits can be viewed as consequences of the assumptions that the instantaneous response and the equilibrium response are elastic.

4. Models.

To get some feeling for linear viscoelastic behavior, it is useful to consider the simpler behavior of analog models constructed from linear springs and dashpots. As analogs for stress and strain, we use the total extending force and the total extension.

The spring is an ideal elastic element obeying the linear force-extension relation $\sigma = G\kappa$. Its relaxation modulus is $G(t) = GH(t)$, and its creep compliance is $J(t) = JH(t)$. Here J is $1/G$.

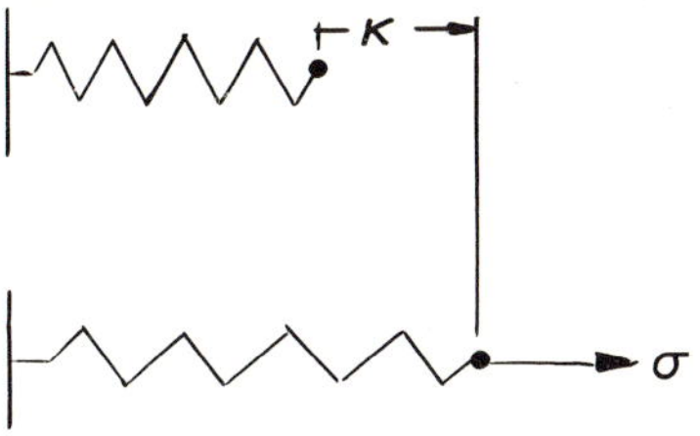

The dashpot is an ideal viscous element that extends at a rate proportional to the applied force, $\dot{\kappa} = \sigma/\eta$. Hence, $J(t) = tH(t)/\eta$ and $G(t) = \eta\delta(t)$.

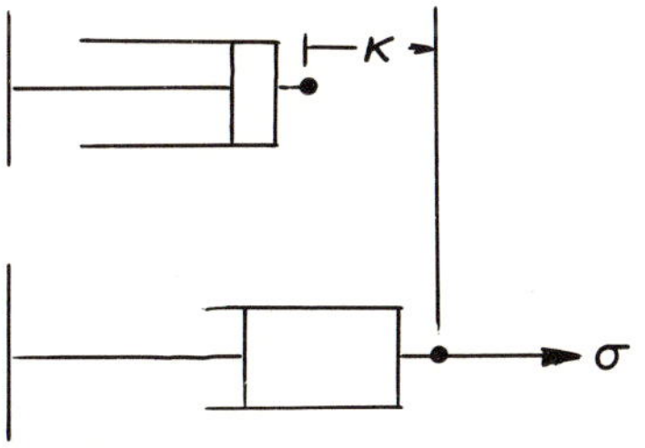

When two elements are combined in series, their compliances are additive. Thus, for example, the Maxwell model consisting of a spring and a dashpot in series has the creep compliance

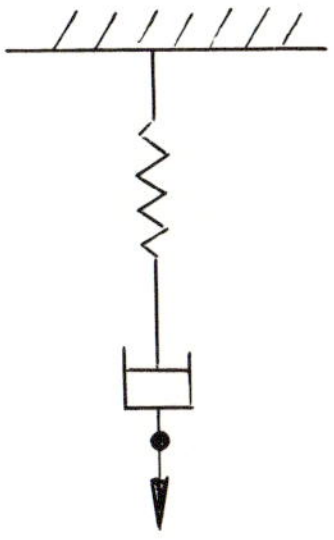

$$J(t) = (J_g + t/\eta)H(t).$$

When two elements are combined in parallel, their moduli are additive. The Kelvin-Voigt model consisting of a spring and dashpot in parallel has the modulus

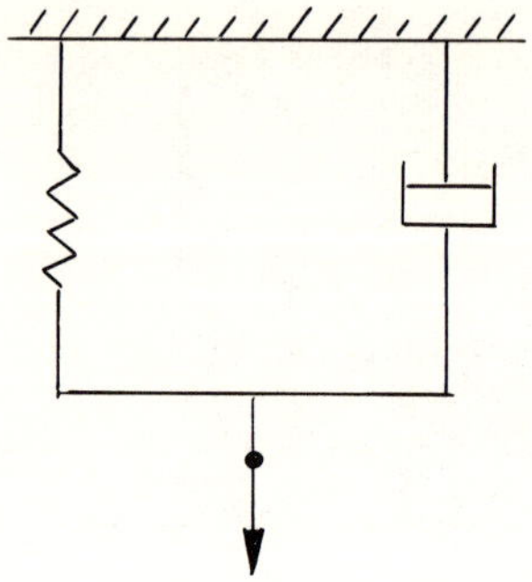

$$G(t) = G_e H(t) + \eta\delta(t).$$

Problems:

1. Find the stress relaxation modulus for a Maxwell model. Does this model represent a solid, or a fluid?

2. Find the creep compliance for a Kelvin-Voigt model. Is this a solid, or a fluid?

3. Find the modulus and compliance of the "standard linear solid", a spring in parallel with a Maxwell model.

5. Superposition.

Knowledge of either one of the single-step response functions $G(t)$ or $J(t)$ is sufficient to allow us to predict the output corresponding to any input, within the linear range in which stresses proportional to κ^3 and strains proportional to σ^3 can be neglected.

First we note that in $G(t)$ and $J(t)$, t is the time lag since application of strain or stress. An input $\kappa(t) = \kappa_o H(t-t_o)$ would be accompanied by an output $\sigma(t) = \kappa_o G(t-t_o)$; we say that the response is time-translation-invariant.

Next consider the stress response to a two-step history,

$$\kappa(t) = H(t-t_1)\Delta\kappa_1 + H(t-t_2)\Delta\kappa_2.$$

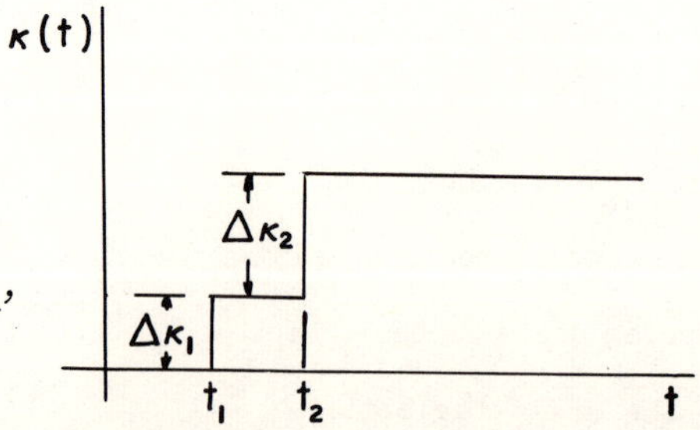

The stress can depend on $t, t_1, t_2, \Delta\kappa_1$, and $\Delta\kappa_2$. We assume that it is a smooth function of the step sizes:

$$\sigma(t) = G_1(t,t_1,t_2)\Delta\kappa_1 + G_2(t,t_1,t_2)\Delta\kappa_2 + O(\Delta\kappa^3).$$

We will neglect the higher-order terms, although it is not essential to the argument to do so.

Since this expression is assumed to be valid for all small-enough step sizes, it holds in particular when $\Delta\kappa_2 = 0$. But in that case, the history is a single-step history, for which the stress response must be $G(t-t_1)\Delta\kappa_1$. Hence $G_1(t,t_1,t_2)$ is identified as $G(t-t_1)$. Similarly, $G_2(t,t_1,t_2)$ must be identical with $G(t-t_2)$. Thus,

$$\kappa(t) = \sum_1^2 H(t-t_\alpha)\Delta\kappa_\alpha \rightarrow \sigma(t) = \sum_1^2 G(t-t_\alpha)\Delta\kappa_\alpha.$$

In the linear approximation, the stress is just the sum of the stresses corresponding to each strain step taken separately. Coupling effects depending on both $\Delta\kappa_1$ and $\Delta\kappa_2$ jointly may occur physically, but in the mathematics they can occur only in higher-order, non-linear approximations.

The same arguments apply equally well to histories with an arbitrary number of steps. The sums over $\alpha = 1,2$, above become sums over $\alpha = 1,2,\ldots N$.

We can approximate any physically realizable strain history arbitrarily well by a step history involving an arbitrarily large number of arbitrarily small steps. We assume that the response of the material is such that when two strain histories are close together, so are the corresponding stress histories. Then the stress corresponding to any given strain history is nearly the same as the stress for a close-by step history. By passing to the limit in the sums above, we obtain

$$\kappa(t) = \int H(t-t')d\kappa(t') \rightarrow \sigma(t) = \int G(t-t')d\kappa(t').$$

Since $G(t)$ is zero for negative t, we can set the upper limit of integration equal to t or ∞ or anything in between, as convenience dictates. The lower limit will usually be written as $-\infty$ in order to be non-committal about when the shearing began, but we implicitly restrict attention to histories with $\kappa = 0$ for all times prior to some starting time, unless and until we find that it makes any sense to do otherwise.

If $\kappa(t)$ has a jump discontinuity $\Delta\kappa_0$ at time t_0, the corresponding contribution to the integral is, of course, $G(t-t_0)\Delta\kappa_0$. Wherever $\kappa(t)$ is differentiable, by $d\kappa(t)$ we mean $\kappa'(t)dt$.

The relation above is the stress-relaxation integral form of the stress-strain relation. All of the preceding remarks hold just as well if we regard the stress history as the input and the strain as the output. We then obtain the creep integral form of the relation:

$$\kappa(t) = \int_{-\infty}^{t} J(t-t')d\sigma(t').$$

The two forms of the relation must be equivalent, of course.

Problem: Verify the following relations:

$$\sigma(t) = \int_{-\infty}^{t} G(t-t')\kappa'(t')dt' = \int_{0}^{\infty} G(t')\kappa'(t-t')dt'$$
$$= (d/dt)\int_{-\infty}^{t} G(t-t')\kappa(t')dt'$$
$$= G_g\kappa(t) + \int_{-\infty}^{t} G'(t-t')\kappa(t')dt'.$$

6. Tensile Response.

We will continue to talk about shearing stress and shearing strain for some time to come, for the sake of definiteness. However, one can obviously consider other kinds of response in the same way. For example, suppose that a rod is extended by the amount $\epsilon(t)$ per unit length by a normal force $\sigma(t)$ per unit area on its ends. Then the extension ϵ and the tensile stress σ are related by expressions of the forms

$$\sigma(t) = \int_{-\infty}^{t} E(t-t')d\epsilon(t') \quad \text{and} \quad \epsilon(t) = \int_{-\infty}^{t} D(t-t')d\sigma(t').$$

Here E and D are respectively the tensile stress relaxation modulus and the tensile creep compliance. Everything that will be said about G and J applies just as well to E and D.

Problems.

1. Suppose that the shear is varied arbitrarily during the interval $(0,t_1)$ and then held constant at the value κ_1 for all times greater than t_1. Show that if G has an equilibrium value G_e, then the stress approaches $\sigma = G_e\kappa_1$ as time increases.

2. Suppose that the stress is varied arbitrarily during the interval $(0,t_1)$ and is then removed, so that $\sigma = 0$ for $t > t_1$. Show that (a) if $J'(\infty) = 0$, the material returns to its initial state after a long time, i.e., $\kappa(\infty) = 0$; but (b) if $J'(\infty) = 1/\eta_0 \neq 0$, recovery is not complete; there is a permanent set equal to the "total viscous flow",

$$\kappa(\infty) = (1/\eta_0) \int_0^{t_1} \sigma(t)dt.$$

3. A constant rate of shear history, $\kappa(t) = \gamma t H(t)$, is applied to a material with modulus $G(t) = 1 + 10^6\exp(-10^6 t) + 10^{-6}\exp(-10^{-6}t)$, $(t > 0)$. Show that if the stress is written in the form

$$\sigma(t) = G\kappa(t) + \eta\kappa'(t),$$

as the sum of an "elastic" and a "viscous" part, then during the interval $(1,10^3)$ the appropriate "constants" are $G = 1.00$ and $\eta = 1.00$, while for $t > 10^7$ the best choices are $G = 1.00$ and $\eta = 2.00$.

4. If $J(t) = t^{\frac{1}{2}}$, guess what $G(t)$ looks like.

7. Relation Between Modulus and Compliance.

Since the creep and relaxation integral formulations must agree with one another, there must be a one-to-one correspondence between modulus and compliance. The basic relation between J and G is found by noticing that if the strain history is $J(t)$, then the stress is $H(t)$, so the stress relaxation integral gives

$$H(t) = \int_{-\infty}^{t} G(t-t')dJ(t') = G(t)J_g + \int_{0+}^{t} G(t-t')J'(t')dt'.$$

If $J_g \neq 0$, we can rearrange this as a Volterra integral equation of the second kind, treating G as the unknown and J as the known function:

$$G(t) = J_g^{-1} - J_g^{-1} \int_0^t J'(t-t')G(t')dt'.$$

Similarly, if G is given and $G_g \neq \infty$, the equation for J is

$$J(t) = G_g^{-1} - G_g^{-1} \int_0^t G'(t-t')J(t')dt'.$$

It is convenient to use the convolution notation,

$$f*g = \int_{-\infty}^{\infty} f(t-t')g(t')dt'.$$

Since G and J are zero for negative values of their arguments, the integrals above are convolutions, and we can write

$$G(t) = G_g H(t) - G_g (J'*G)(t)$$

and

$$J(t) = J_g H(t) - J_g (G'*J)(t).$$

Problem: Prove that $f*g = g*f$ and $(f*g)*h = f*(g*h)$.

These equations can be solved to any desired degree of accuracy by iteration. For example, if J is given then to find G, we begin with any estimate G_1, even a very bad one, and then form the sequence of functions G_n by the rule

$$G_{n+1} = G_g H - G_g J'*G_n.$$

It is a standard result in the theory of integral equations that the sequence $G_n(t)$ converges to $G(t)$, for each t. A solution exists, and it is unique. Convergence of the iteration process is not uniform in t, however. The larger t is, the more iterations are required before a reasonable approximation to $G(t)$ is obtained. Consequently, this iterative method is usually not a practical method of solution except for small values of t.

Taking it for granted that J is an increasing function and G is decreasing (for $t > 0$), the integral equation yields

$$1 = \int_{-\infty}^{t} G(t-t')dJ(t') \geq G(t) \int_{-\infty}^{t} dJ(t') = G(t)J(t).$$

Thus,

$$\boxed{J(t)G(t) \leq 1.}$$

Under what circumstances can the inequality be an equality?

Problems.

1. Assume that $G_g > 0$ and $G' < 0$. Show that in determining J by iteration, every J_n is positive if J_1 is, and thus that J is positive.

2. The result of iteration can be expressed as an infinite series,

$$G = G_g H - G_g^2 J'*H + G_g^3 J'*J'*H - G_g^4 J'*J'*J'*H + \cdots .$$

Show why. Given $J(t) = 2 - \exp(-t/T')$, evaluate $(J'*)^n$, the result of folding J' with itself n times. Sum the series to find G. (The object is not to find G, which is easy, but to look at the convergence of the iteration procedure.)

3. In the preceding example, show that the partial sums of the infinite series are alternately upper and lower bounds on $G(t)$. Plot graphs of the first four partial sums.

4. Show that for t positive,

$$\int_0^t J(t-t')G(t')dt' = t.$$

5. Prove the following inequalities:

$$G(t) \int_0^t J(t')dt' \leq t \leq J(t) \int_0^t G(t')dt'.$$

One of these is not as close as $J(t)G(t) \leq 1$; the other gives new information. Explain.

8. Sinusoidal Shearing.

The properties of viscoelastic materials can be described not only in terms of the response to a step loading or a step deformation, but also in terms of the responses to sinusoidal inputs.

If we say that the shear is given for all time by $\kappa(t) = \kappa_0 \exp(i\omega t)$ (real part understood), a sinusoidal shearing with amplitude κ_0 and radian frequency ω, and we try to evaluate the stress by using the stress-relaxation integral, we find that in general the integral does not converge. To get a sensible result, we must confess that the shearing has not been going on forever, but started at some definite time, zero, say. There is then no trouble about divergent integrals. It is also a convenience to write the relaxation integral as

$$\sigma(t) = G(t)\kappa(t) + \int_0^\infty [G(t') - G(t)]\kappa'(t-t')dt'.$$

Then, for the oscillatory history starting at time zero, we obtain

$$\sigma(t) = G(t)\kappa_0 \exp(i\omega t) + \kappa_0 i\omega \exp(i\omega t) \int_0^t [G(t') - G(t)]\exp(-i\omega t')dt'.$$

We assume that G approaches a limit G_e, fast enough for the limit of the integral to exist. Then for t approaching infinity (a few relaxation times), the stress approaches a sinusoidal oscillation,

$$\sigma(t) = G^*(\omega)\kappa_0 \exp(i\omega t).$$

The proportionality constant G^*, which is called the dynamic modulus or complex modulus, is the complex number defined for each ω by

$$G^*(\omega) = G_e + i\omega \int_0^\infty [G(t) - G_e]\exp(-i\omega t)dt.$$

Later an easier way to compute G^* from G will be given.

The relation between stress and strain in sinusoidal shearing can also be written as $\kappa(t) = J^*(\omega)\sigma(t)$, where J^*, the complex compliance, is the reciprocal of G^*:

$$\boxed{J^*(\omega)G^*(\omega) = 1.}$$

If J approaches a limit J_e (i.e., the material is a solid), then J^* can be expressed in terms of J in the same way that G^* is expressed in terms of G. However, if the material is a fluid, J increases without bound as t increases, and no mere rearrangement can produce a convergent integral.

Problem: Show that if a fluid is subjected to a stress history $\sigma(t) = H(t)\sin(\omega t)$, then after a long time the shear oscillates sinusoidally about a non-zero value. Explain why in physical terms.

9. Nomenclature.

If J^* and G^* are written in polar form as

$$J^* = |J^*|\exp(-i\delta), \quad G^* = |G^*|\exp(i\delta),$$

then $|G^*|$ is the ratio of the stress amplitude to the shear amplitude, and the loss angle δ represents the phase shift between input and output. It is found empirically that the loss angle always lies between zero and $\pi/2$. For hard solids, it is small, so that the stress and strain are almost in phase.

If G^* and J^* are separated into real and imaginary parts as

$$G^* = G_1 + iG_2, \quad J^* = J_1 - iJ_2,$$

then G_1 is called the storage modulus and G_2 is the loss modulus, with similar terms for the compliance. The loss tangent is the tangent of the loss angle:

$$\tan \delta(\omega) = \frac{G_2(\omega)}{G_1(\omega)} = \frac{J_2(\omega)}{J_1(\omega)} .$$

The quantity $G_2(\omega)/\omega$ is the dynamic viscosity. $G^*(\omega)/i\omega$ is the complex viscosity. It is used more in connection with fluids, and is just the Fourier transform of G (for fluids). The notation $G' + iG''$ is usually used instead of $G_1 + iG_2$, but we reserve primes to mean differentiation.

10. Energy Storage and Loss.

The names, storage modulus and loss modulus, connote something to do with energy storage and loss. Notice that with $\kappa(t) = \kappa_o \exp(i\omega t)$, then $\kappa'(t) = i\omega\kappa(t)$, so the stress can be written as

$$\begin{aligned} \sigma(t) &= G^*\kappa(t) = (G_1+iG_2)\kappa(t) \\ &= G_1\kappa(t) + (G_2/\omega)\kappa'(t). \end{aligned}$$

The stress is thus represented as an elastic part with modulus G_1 plus a viscous part with viscosity G_2/ω. (Thus a Kelvin-Voigt model fits the behavior at any particular frequency, but the coefficients for the model must be changed every time the frequency is changed.)

The work done by the stress on the change of strain is then

$$dW = \sigma d\kappa = d(\tfrac{1}{2} G_1\kappa^2) + (G_2/\omega)(\kappa')^2 dt.$$

The part associated with G_1 is a perfect differential, and can be interpreted as a change of an elastic stored energy. The part associated with G_2 is like a rate of viscous dissipation. The total work in a complete cycle is

$$(G_2/\omega) \oint (\kappa')^2 dt.$$

Since you don't get something for nothing (first law of thermodynamics) and you don't even get nothing for nothing (second law), it requires positive power to deform a material sinusoidally, so G_2 is positive at all positive frequencies. Oddly enough, there is no purely thermodynamical reason why the storage modulus G_1 should be positive, but it always is.

11. Oscillation with Increasing Amplitude.

Although it is an un-physical kind of thing to consider, it turns out to be profitable to look at cases involving a shear oscillating with an exponentially growing amplitude:

$$\kappa(t) = \kappa_0 \exp(rt+i\omega t) = \kappa_0 \exp(st), \quad s = r + i\omega.$$

The amplitude of oscillation is $\kappa_0 \exp(rt)$, so it is extremely small in the distant past, and we can consider that the same history has been going on forever without getting into trouble with divergent integrals. The stress-relaxation integral gives

$$\sigma(t) = \int_0^\infty G(t')\kappa_0 s \exp(s(t-t'))dt' = s\overline{G}(s)\kappa_0 \exp(st).$$

We see that the stress also oscillates at frequency ω with an amplitude growing in proportion to $\exp(rt)$. The proportionality constant, a complex number, is given by

$$s\overline{G}(s) = s \int_0^\infty G(t)\exp(-st)dt.$$

The function $\overline{G}(s)$ defined by the integral is the Laplace transform of G. We assume that the integral is convergent whenever r, the real part of s, is positive, so that the distant past cannot contribute much to the stress.

Equally well, we can start out by supposing that the stress has been oscillating with exponentially growing amplitude forever, so that $\sigma(t) = \sigma_0 \exp(st)$. Then computing the shear from the creep integral, we obtain

$$\kappa(t) = s\overline{J}(s)\sigma_0 \exp(st),$$

where

$$s\bar{J}(s) = s\int_0^\infty J(t)\exp(-st)dt.$$

Again, we assume that the integral converges for all positive r.

Of course, these are two descriptions of one process, in which both stress and strain grow in proportion to $\exp(st)$. Since the two descriptions must agree, we find by comparing them that $s\bar{G}$ and $s\bar{J}$ must be reciprocal:

$$\boxed{s\bar{G}(s)s\bar{J}(s) = 1.}$$

This relation between the transforms allows us to determine G, given J, or vice versa, in very simple cases. More importantly, even in realistic cases in which the transforms are not easy to compute, the relation gives a simple way to establish connections between gross features of G and J, as we shall see later.

For a constant-amplitude shearing history $(r = 0)$, the integral defining $s\bar{G}(s)$ is not necessarily convergent. However, we assume that its limit as $s\ (= r + i\omega)$ approaches $i\omega$ does exist. By comparing this limit with the expression for G^* we find that the dynamic modulus is given in terms of the transform by

$$G^*(\omega) = \lim_{r \to 0} (r+i\omega)\bar{G}(r+i\omega).$$

Ordinarily, no limiting process is actually involved. We simply compute $s\bar{G}(s)$ and then evaluate its analytic continuation on the pure imaginary axis.

Similarly, $J^*(\omega)$ is evaluated by computing $s\bar{J}(s)$ and then setting $s = i\omega$ in the expression for its analytic continuation.

Examples: For an elastic solid, $G = G_e$. Since the Laplace transform of 1 is $1/s$ (verify), then $s\bar{G}(s) = G_e$ as well. Since $s\bar{J}$ is the reciprocal of $s\bar{G}$, then $s\bar{J}(s) = 1/G_e = J_e$. Hence, $J(t) = J_e$, as we know. Since the complex modulus is obtained by setting $s = i\omega$ in the expression for $s\bar{G}$, we get $G^*(\omega) = G_e$ also, as expected, and then $J^* = 1/G^* = 1/G_e = J_e$. The storage modulus is $G_1 = G_e$ and the loss modulus is $G_2 = 0$. The loss angle and the loss tangent are zero. The dynamic viscosity G_2/ω is zero.

For a viscous fluid, $J(t) = t/\eta_o$. Since the Laplace transform of t is $1/s^2$, then $s\bar{J}(s) = s^{-1}/\eta_o$. Hence, $s\bar{G}(s) = \eta_o s$. The transform of $\delta(t)$ is unity, so $G(t) = \eta_o\delta(t)$. The complex modulus and compliance are $G^* = \eta_o i\omega$ and $J^* = -i\eta_o/\omega$. The loss angle is $\pi/2$; the stress leads the strain by $\pi/2$ radians because it is in phase with the strain-rate. The loss tangent is infinite. The dynamic viscosity G_2/ω is η_o. The storage modulus G_1 is zero.

The compliance of a Maxwell model is $J(t) = (T+t)/\eta_o$, where $T = J_g\eta_o$. Hence $s\bar{J}(s) = (T+s^{-1})/\eta_o$, and $s\bar{G}(s) = \eta_o s/(sT+1)$, or, with $\eta_o = G_gT$, $s\bar{G}(s) = G_g s/(s+T^{-1})$. Since $1/(s+c)$ is the transform of $\exp(-ct)$, (verify), then $G(t) = G_g\exp(-t/T)$. Notice that the viscosity η_o is the integral of G from zero to infinity. The complex modulus is

$$G^*(\omega) = G_g \frac{i\omega T}{1 + i\omega T} = G_g \frac{(\omega T)^2 + i\omega T}{1 + (\omega T)^2} .$$

We see that G^* approaches G_g at high frequencies. G_1 approaches G_g and G_2 approaches zero. The material behaves like an elastic material with modulus G_g if the frequency is high. At low frequencies, the dynamic modulus is asymptotic to $i\omega G_gT = i\omega\eta_o$, as for a viscous fluid. The loss tangent is $1/\omega T$.

CHAPTER II

FOURIER AND LAPLACE TRANSFORMS

The simplest methods of determining J, given G, or vice versa, are based on the use of Laplace transforms. Fourier and Laplace transforms also find a variety of other applications in connection with viscoelasticity theory. Let us briefly review these transform methods.

1. Fourier Transforms.

The Fourier series representation of a function f defined on the interval $(-L,L)$ has the form

$$f(t) = (2\pi)^{-1} \sum_{-\infty}^{\infty} f_n \exp(in\pi t/L).$$

The complex-looking form of the series is always more convenient than sines and cosines for analytical manipulations. The coefficients f_n are found by multiplying both sides of the equation by $\exp(-in\pi t/L)$ and integrating from $-L$ to L. This yields

$$\int_{-L}^{L} f(t)\exp(-in\pi t/L)dt = Lf_n/\pi.$$

If we write $\omega_n = n\pi/L$, $\Delta\omega_n = \pi/L$, and $\bar{f}(\omega_n) = Lf_n/\pi$, then the series is

$$f(t) = (2\pi)^{-1} \sum_{-\infty}^{\infty} \bar{f}(\omega_n)\exp(i\omega_n t)\Delta\omega_n,$$

with

$$\bar{f}(\omega) = \int_{-L}^{L} f(t)\exp(-i\omega t)dt.$$

When L is large, the frequencies ω_n are very close together, and the series is a Riemann sum approximating an integral. Passing to the limit as $L \to \infty$, we get

$$f(t) = (2\pi)^{-1} \int_{-\infty}^{\infty} \bar{f}(\omega)\exp(i\omega t)d\omega,$$

where

$$\bar{f}(\omega) = \int_{-\infty}^{\infty} f(t)\exp(-i\omega t)dt.$$

Thus $f(t)$ is decomposed into sinusoids, the amplitude of the component at frequency ω being $\bar{f}(\omega)/2\pi$. $\bar{f}(\omega)$ is the Fourier transform of $f(t)$. The advantage of this decomposition is that linear operations on $f(t)$ can now be carried out just as if f were itself sinusoidal. Differentiating or integrating a sinusoid (i.e., complex exponential) just multiplies the amplitude by a constant, so calculus is changed into algebra.

2. Two-Sided Laplace Transforms.

The fly in the ointment is that most simple functions do not have a Fourier transform, because the defining integral fails to converge at its infinite limits. This difficulty can sometimes be overcome by considering not $f(t)$ but $f(t)\exp(-rt)$. The transform of the latter, if it exists, is

$$\int_{-\infty}^{\infty} f(t)\exp(-st)dt \equiv \bar{f}(s),$$

where $s = r + i\omega$. By multiplying the inversion formula on both sides by $\exp(rt)$, we obtain

$$f(t) = (2\pi)^{-1} \int_{-\infty}^{\infty} \bar{f}(r+i\omega)\exp[(r+i\omega)t]d\omega.$$

The function $\bar{f}(s)$ is the two-sided Laplace transform of $f(t)$.

Often convergence at the upper limit can be produced by taking r sufficiently large, but the damping factor $\exp(-rt)$ only makes matters worse at the lower limit.

3. Laplace Transforms.

However, if we are really only interested in values of $f(t)$ for t positive, it is sufficient to consider the function $H(t)f(t)$, where $H(t)$ is the Heaviside unit step function. One side of the two-sided transform is cut off, and convergence on the other side is achieved by using the damping factor:

$$\overline{f}(s) = \int_0^\infty \exp(-st)f(t)dt.$$

This is what is called the Laplace transform. It is the Fourier transform of a chopped and damped version of the function f. The inversion integral is

$$H(t)f(t) = \frac{1}{2\pi} \int_{-\infty}^{\infty} \exp[(r+i\omega)t]\overline{f}(r+i\omega)d\omega = \frac{1}{2\pi i} \int_{r-i\infty}^{r+i\infty} \exp(st)\overline{f}(s)ds.$$

The first integral defines what is meant by the second one.

If the integral defining the Laplace transform converges for some damping factor $r = \mathrm{Re}(s)$, it also converges (absolutely) for all larger damping factors, because a larger damping makes the integrand smaller at every point. Thus, in the complex s-plane, the integral converges in some half-plane $\mathrm{Re}(s) > r_o$ ($r_o = \pm\infty$ not excluded), called the half-plane of convergence. The value r_o at the boundary of this half-plane is the abscissa of convergence. It divides the values of r that are big enough from those that aren't. r_o itself may or may not be big enough.

The integral converges uniformly at its upper limit for all s in any compact region inside the half-plane of convergence. This incantation wards off evil spirits if we should want to interchange orders of integration, which we do want to do now: By integrating $\overline{f}(s)$ around a closed contour, we get zero because the integral of $\exp(-st)$ around a closed contour is zero. Thus $\overline{f}(s)$ is an analytic function of s in the half-plane of convergence. Accordingly, it can be extended by analytic continuation to values of s not in the half-plane of convergence. It is the complete function rather than the integral that is called the transform.

The function $\bar{f}(s)$ has no singularity in the half-plane of convergence, obviously, but we should expect that it would have a singularity for some s on the abscissa of convergence, and it does (but not obviously). Consequently, given $\bar{f}(s)$, we can identify r_0 by locating the singularity of $\bar{f}(s)$ with greatest real part. This is important because the inversion integral is an integral along a line $Re(s) = r = $ constant with $r > r_0$. Thus, the inversion contour must pass to the right of all singularities of $\bar{f}(s)$.

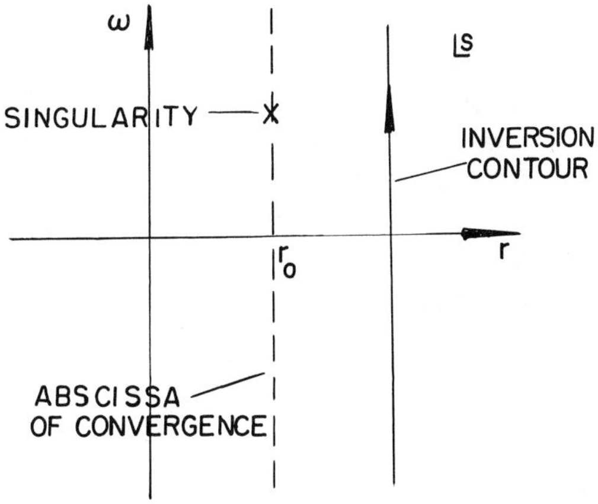

The analyticity of $\bar{f}(s)$ makes life easier in many ways. For example, let's compute the transform of t^n. First suppose that s is a large, real, positive number. Then

$$\int_0^\infty \exp(-st)t^n dt = s^{-n-1} \int_0^\infty \exp(-x)x^n dx,$$

where we have made the change of variable $st = x$. The new integral is equal to $n!$, so the transform of t^n is $n!s^{-n-1}$, at least if s is large, real, and positive. But the analytic continuation of this function to other values of s is trivial; the notation takes care of it.

By the way, for all values of z with $Re(z) > -1$, we define $z!$ by the integral

$$z! = \int_0^\infty \exp(-x)x^z dx.$$

Problems.

1. Prove by induction that if z is a positive integer, the integral above is equal to $1\cdot 2\cdot 3\cdot \ldots \cdot(z-1)z$.

2. Show that even if z is not an integer, $z! = z(z-1)!$

3. Separate z into real and imaginary parts and then separate the integral into real and imaginary parts. Show why the restriction $\mathrm{Re}(z) > -1$ is needed.

4. Show that the transform of t^p is $p!s^{-p-1}$ if $\mathrm{Re}(p) > -1$, and that otherwise t^p has no transform.

5. What is the half-plane of convergence for the Laplace integral of t^p? What is the abscissa of convergence? Where are the singularities of the transform, and what kind are they (depending on what p is)?

4. Elementary Formulas.

We use the notation $L(f)$ to mean the same thing as $\bar{f}(s)$. Some of the more usual manipulations of Laplace integrals are conveniently summarized as formulas. The formulas are valid only when all quantities in them are well-defined.

Integration by parts gives

$$L(f') = s\bar{f}(s) - f(0+).$$

Then by recursion,

$$L(f^{(n)}) = s^n\bar{f}(s) - s^{n-1}f(0+) - \dots - f^{(n-1)}(0+).$$

By differentiating $L(f)$ with respect to s we obtain

$$\bar{f}^{(n)}(s) = (-1)^n L(t^n f).$$

The scaling formula for either f or $\bar{f}$ is

$$L[f(t/c)] = c\bar{f}(cs) \qquad (c > 0).$$

Shifting a function to the right decreases its transform exponentially:

$$L[H(t-t_0)f(t-t_0)] = \bar{f}(s)\exp(-st_0).$$

Damping the function shifts its transform:

$$L[\exp(-ct)f(t)] = \bar{f}(s+c).$$

The shift is toward the left if $c > 0$, making the half-plane of convergence larger.

If we define

$$f_{n+1}(t) = \int_0^t f_n(t')dt',$$

with $f = f_0$, then f is the nth derivative of f_n, and all lower derivatives of f_n have initial value zero. Then the derivative formula gives

$$L(f_n) = s^{-n}\bar{f}(s).$$

The analogous formula for the integral of the transform is

$$\int_s^\infty \bar{f}(s')ds' = L(t^{-1}f).$$

This formula usually doesn't make sense.

<u>Example</u>: As an example of the use of Laplace transforms, let us evaluate the integral

$$\int_0^\infty t^{-1}[\exp(-at) - \exp(-bt)]dt.$$

Although the integrand is not singular at $t = 0$, both terms of it are, a frustrating situation. The easiest way to do the integration is to observe that it (or any integral from 0 to ∞) is the value at $s = 0$ of the transform of the integrand. First, the transform of 1 is $1/s$; then from the shift formula, the transform of exp(-at) is 1/(s+a); then from the integral formula, the transform of the integrand is

$$\int_s^\infty \left(\frac{1}{s' + a} - \frac{1}{s' + b}\right) ds' = \log \frac{s + b}{s + a} .$$

Evaluating this at $s = 0$, we find that the original integral is equal to log(b/a). (We use log to mean the natural logarithm.)

Problems.

1. Verify all of the formulas that are not familiar.

2. Use the formulas, rather than direct integration, to evaluate the transforms of each of the following functions in sequence, starting with the result that the transform of the Dirac delta $\delta(t)$ is unity. (a) H(t). (b) exp(-ct). (c) $\cos(ct) = \frac{1}{2}(e^{ict}+e^{-ict})$. (d) sin(ct) (from derivative of cosine). (e) $t^{-1}\sin(t)$.

3. Use the result of problem 2(e) to evaluate the integral of $t^{-1}\sin(t)$ from zero to infinity.

4. Evaluate the transform of $t^p\exp(-t/T)$.

5. Why do the formulas come in pairs?

5. Convolutions.

If f and g vanish for negative argument, their convolution is

$$f*g = \int_{-\infty}^{\infty} f(t-t')g(t')dt' = \int_0^t f(t-t')g(t')dt' .$$

In evaluating the transform of the convolution it simplifies interchange of the order of integration to leave the limits infinite. Then

$$L(f*g) = \int\int_{-\infty}^{\infty} \exp(-st)f(t-t')g(t')dt\, dt' .$$

The interchange of order of integration is either valid or not, and either way it is nothing to worry about. Substitute $x = t - t'$ for t as an integration variable, and get

$$L(f*g) = \iint_{-\infty}^{\infty} \exp(-sx)f(x)\exp(-st')g(t')dx\,dt'$$
$$= \bar{f}(s)\bar{g}(s).$$

Similarly, the transform of fg is the convolution of $\bar{f}$ with $\bar{g}$ over an appropriate contour, but this result is not of much use.

Since multiplication of transforms is commutative and associative, it follows that so is convolution. This means that in a convolution of several functions, $f*g* \cdots *h$, the integrations can be carried out in whatever order is most convenient. For example, to evaluate the nth integral of f, $f_n = H*H* \cdots *H*f$, it is easiest to evaluate $H*H* \cdots *H$ first. It is equal to

$$(H*)^n = t^{n-1}/(n-1)!$$

(<u>Problem</u>: Prove this.) Hence

$$\int_0^t \int_0^{t_n} \cdots \int_0^{t_2} f(t_1)dt_1 dt_2 \cdots dt_n = \int_0^t \frac{(t-t')^{n-1}}{(n-1)!} f(t')dt'.$$

We make heavy use of transforms in solving convolution integral equations. Consider the following example:

$$\int_0^x \frac{u(y)dy}{\sqrt{x-y}} = e^{x^2} - 1 \equiv f(x) \quad \text{(say)}.$$

To solve for u, we take the transform of both sides, and get

$$\bar{u}(s)(-\tfrac{1}{2})!/s^{\frac{1}{2}} = \bar{f}(s).$$

Then

$$\bar{u}(s) = \frac{1}{[(-\frac{1}{2})!]^2} \frac{(-\frac{1}{2})!}{s^{\frac{1}{2}}} s\bar{f}(s).$$

Here $(-\frac{1}{2})! = \sqrt{\pi}$. Since $f(0) = 0$, $s\bar{f}(s)$ is the transform of the derivative of f, $f'(x) = 2x \exp(x^2)$. We again use the convolution theorem to invert the right-hand side, and get

$$u(x) = \frac{2}{\pi} \int_0^x \frac{y e^{y^2} dy}{\sqrt{x-y}} .$$

Notice that we never needed to calculate the transform of $f(x)$. Notice that we would have been in hot water if we had tried to calculate it, because it doesn't have a transform. Nevertheless, everything is perfectly all right.

Problem: Does $H(10^{10}-x)\exp(x^2)$ have a transform? Replacing $\exp(x^2)$ by this in the problem above, what is the solution for $x < 10^{10}$?

Problem: Use the convolution theorem to obtain the relation between $s\bar{G}$ and $s\bar{J}$ from the equation

$$1 = G(t)J_g + \int_{0+}^{t} G(t-t')J'(t')dt' .$$

Example: If the integrand of a definite integral is zero or infinite at each limit of integration, it is worth trying to find a change of variable that will put it into the following form:

$$\int_0^1 t^p(1-t)^q dt \equiv B(p+1, q+1).$$

$B(p,q)$ is called a Beta function. To evaluate the integral, notice that it is the value at $x = 1$ of the convolution

$$f(x) = \int_0^x (x-y)^q y^p dy.$$

Then

$$\bar{f}(s) = \frac{q!}{s^{q+1}}\,\frac{p!}{s^{p+1}} = \frac{p!q!}{(p+q+1)!}\,\frac{(p+q+1)!}{s^{p+q+1+1}} .$$

Then inverting and setting $x = 1$, we get

$$\int_0^1 t^p(1-t)^q dt = \frac{p!q!}{(p+q+1)!} .$$

Problem: Show that when $p = q = \frac{1}{2}$, the graph of the integrand is a semicircle. What is its area? What is the value of $\frac{1}{2}!$?

Example: When $q = -p$, the change of variables $x = t/(1-t)$ gives

$$p!(-p)! = \int_0^1 \left(\frac{t}{1-t}\right)^p dt = \int_0^\infty \frac{x^p}{(x+1)^2}\,dx.$$

With p not an integer in general (for what values of p is the integral convergent?), we introduce a branch cut along the positive real axis and then integrate over the contour shown. By calculating the residue at $x = -1$ and performing some manipulations, we get

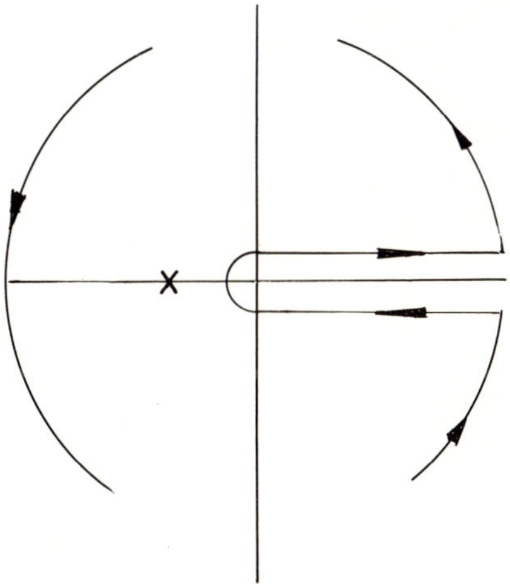

$$p!(-p)! = \frac{p\pi}{\sin p\pi} .$$

This is the Reflection Formula.

Problem: Carry out the details.

Example: If $G(t)$ is a relaxation modulus then, so far as anyone knows, $G(0+)$ is finite. (It is hard to measure.) Nevertheless, it is sometimes useful to use approximations such as $G(t) = G_0 t^{-p}$ $(0 \le p < 1)$ for the behavior near some point $t = t_0$. The approximation is infinite at $t = 0$, but this doesn't cause trouble because the singularity is integrable. The s-multiplied transform is

$$s\bar{G}(s) = G_o(-p)!s^p.$$

Earlier we found by a direct argument that $s\bar{J}$ is the reciprocal of $s\bar{G}$. We could also derive this from the integral relation between J and G by using the convolution theorem. We obtain

$$s\bar{J}(s) = \frac{p!s^{-p}}{G_o(-p!)p!} .$$

Since $p!(-p)!$ is equal to $p\pi$ over $\sin(p\pi)$, then

$$J(t) = J_o t^p,$$

where

$$J_o G_o = \frac{\sin(p\pi)}{p\pi} .$$

If p is close to zero, which is usually the case when we use this sort of approximation, $\sin(p\pi)/(p\pi)$ is nearly unity, so J is nearly the reciprocal of G. As we know, JG must be less than unity, but it is only a little less if p is small.

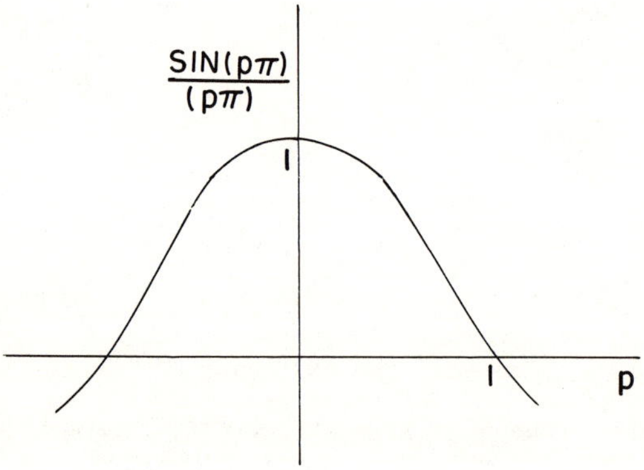

CHAPTER III

RELATIONS BETWEEN MODULUS AND COMPLIANCE

Connections between some of the gross features of a modulus and the corresponding compliance are easy to obtain by using the reciprocal relation between their s-multiplied transforms. In the present chapter we first consider some exact relations that can be obtained by considering the behavior of the transforms for small values of the transform parameter. We then consider some approximate but more detailed relations that can be obtained by using linear approximations to the response functions on a log-log plot. The main object of these considerations is to develop enough qualitative understanding that, given one response function in graphical or numerical form, graphs of all others can be sketched immediately with fair quantitative accuracy. More refined numerical techniques which can give arbitrarily accurate results are not discussed because they do not lead to any qualitative understanding of the situation.

1. Limits and Moments: Fluids.

The relaxation modulus for a fluid has an equilibrium value $G_e = 0$. We suppose that G approaches zero fast enough that its integral from zero to infinity converges:

$$\eta_o = \int_0^\infty G(t)dt.$$

This is a consequence of the assumption that in steady shearing starting at time zero, the stress approaches a limiting value as time increases. Calling the limit $\gamma\eta_o$, where γ is the shear rate, we find that the viscosity coefficient η_o is the integral of G.

We also assume that the integral defining the first moment of G is convergent:

$$\eta_o T = \int_0^\infty tG(t)dt.$$

T, the centroid of $G(t)$, is the <u>mean relaxation time</u>. We would also be willing to assume that higher moments exist, if the assumption should prove useful:

$$\int_0^\infty t^n G(t)dt < \infty.$$

From the integral defining the transform of G,

$$\overline{G}(s) = \int_0^\infty \exp(-st)G(t)dt,$$

we see that $\overline{G}(0)$ is the steady-shearing viscosity η_0. When s is small, by expanding the exponential in powers of s we find that

$$\overline{G}(s) = \eta_0[1 - Ts + \cdots].$$

The coefficients of the higher powers of s are the higher moments of G, so far as these exist. If the moments fail to exist beyond some stage, then the terms that do make sense give the asymptotic behavior near $s = 0$.

This series determines the behavior of the dynamic modulus at low frequencies:

$$G^*(\omega) = i\omega\overline{G}(i\omega) = i\omega\eta_0[1 - Ti\omega + \cdots].$$

Thus the storage modulus G_1 is asymptotic to $\eta_0 T\omega^2$, and the loss modulus G_2 is asymptotic to $\eta_0\omega$; the dynamic viscosity G_2/ω approaches the steady-shearing viscosity η_0 at low frequencies. Data on these quantities are usually graphed on log-log plots. For $\log \omega \to -\infty$, we have $\log G_1 \sim 2 \log \omega + \log(\eta_0 T)$ and $\log G_2 \sim \log \omega + \log \eta_0$.

The complex compliance, $1/G^*$, becomes infinite at zero frequency because the deformation of a fluid under a steady stress becomes infinitely large:

$$J^*(\omega) = \frac{1}{i\omega\eta_o} + \frac{T}{\eta_o} + \cdots .$$

The transform of J has a pole at $s = 0$:

$$\bar{J}(s) = \frac{1}{\eta_o}\left[\frac{1}{s^2} + \frac{T}{s} + \cdots\right].$$

The formal inverse of this is

$$J(t) = (1/\eta_o)(t + T + \cdots).$$

This represents the behavior for large values of t. The terms indicated by dots represent a function that approaches zero at large t, whose transform is regular at $s = 0$.

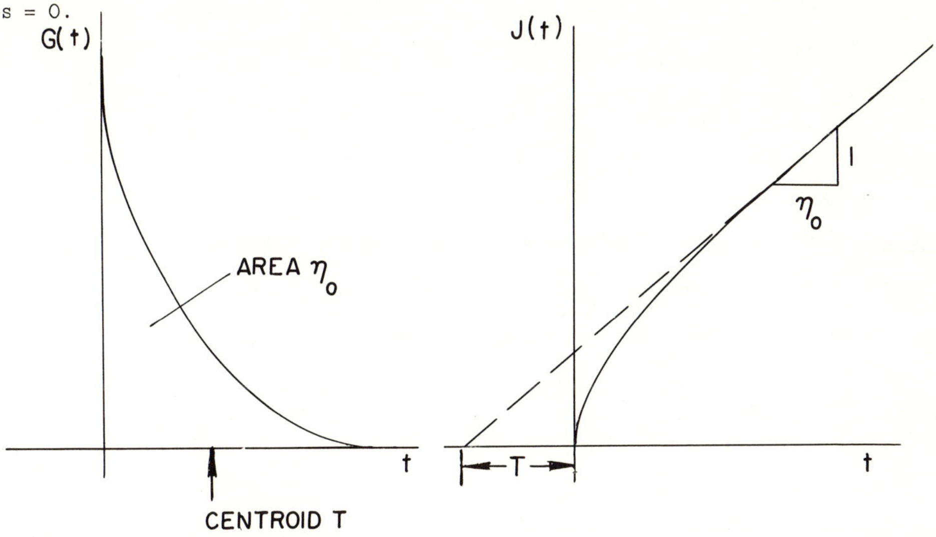

2. Limits and Moments: Solids.

We call a material a solid if the integral of G from zero to infinity diverges. This includes those cases in which the equilibrium modulus G_e is not zero. It also includes cases, probably un-physical, in which $G_e = 0$ but the approach to the limit is not fast enough for integrability. For example, if G should behave like $1/t$ for large t, then the stress would grow infinitely large in a steady-shearing history, and we would call the material a solid.

Let us suppose that G_e is not zero and that the approach to the equilibrium value is exponentially fast. We can write the s-multiplied transform as

$$s\bar{G}(s) = G_e + s\int_0^\infty \exp(-st)[G(t) - G_e]dt.$$

For small values of s this yields

$$s\bar{G}(s) = G_e + \eta_o s - \eta_o T s^2 + \cdots .$$

Here, η_o and T are defined by

$$\eta_o = \int_0^\infty [G(t) - G_e]dt$$

and

$$\eta_o T = \int_0^\infty t[G(t) - G_e]dt.$$

Problem: Show that with steady shearing beginning at time zero, the stress is asymptotic at large times to $\sigma(t) \sim G_e\kappa(t) + \eta_o\kappa'(t)$.

The dynamic modulus of a solid is given at low frequencies by

$$G^*(\omega) = G_e + \eta_o i\omega - \eta_o T(i\omega)^2 + \cdots .$$

Thus, the storage modulus G_1 approaches the equilibrium modulus G_e at zero frequency, and the dynamic viscosity G_2/ω approaches η_o. The complex compliance is

$$J^*(\omega) = J_e[1 - J_e\eta_o i\omega + J_e\eta_o(T+J_e\eta_o)(i\omega)^2 - \cdots],$$

where we have written J_e for $1/G_e$. The transform of J is

$$s\bar{J}(s) = J_e[1 - J_e\eta_o s + J_e\eta_o(T+J_e\eta_o)s^2 - \cdots].$$

The coefficients in this expansion can be interpreted in the same sort of way as the coefficients in the expansion of $s\bar{G}$. The first term, J_e, is the limiting value of J. The next coefficient, $-J_e^2\eta_o$, is the integral of $J - J_e$ from zero to infinity. Thus, recalling how η_o is defined, we find that

$$\frac{1}{J_e}\int_0^\infty [J_e - J(t)]dt = \frac{1}{G_e}\int_0^\infty [G(t) - G_e]dt.$$

The third coefficient represents the integral of $t(J-J_e)$. We define the <u>mean retardation time T'</u> by

$$T' = \int_0^\infty t[J_e - J(t)]dt/\int_0^\infty [J_e - J(t)]dt.$$

Then from the third coefficient we find that

$$T' = T + \eta_o/G_e.$$

Evidently the mean retardation time is always greater than the mean relaxation time.

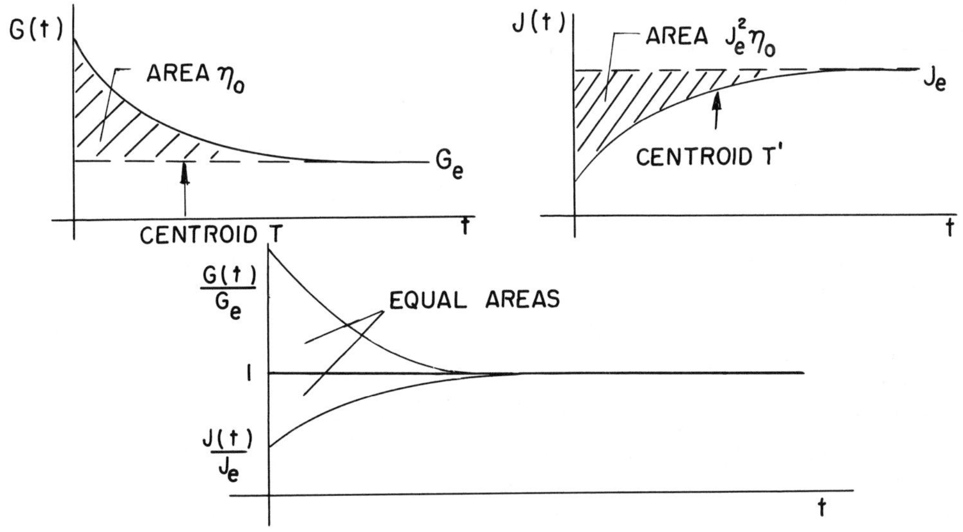

Problems.

1. For the modulus $G = G_e + (G_g - G_e)\exp(-t/T)$, evaluate η_o and show that T is the mean relaxation time. Assuming that the compliance has the form $J = a - b\exp(-t/T')$, use the values of G_g, G_e, η_o, and T to evaluate a, b, and T'. Then verify that $s\bar{J}$ and $s\bar{G}$ are reciprocal.

2. For the modulus $G = G_o t^{-p}$ and corresponding compliance $J = J_o t^p$, the various limits and moments that have been discussed do not exist. However, if p is small, J and G will have levelled off enough by time t_o, say, that one might become convinced by looking at graphs of these functions that equilibrium values $G_e = G_o t_o^{-p}$ and $J_e = J_o t_o^p$ had been reached. Find the values of η_o and T that would be obtained from G under this false assumption, and the values of η_o and T' that would be obtained from J. Show that the two values of η_o are nearly equal, and that the general relation between T and T' is approximately satisfied.

3. For an arbitrary modulus $G(t)$, define an apparent viscosity $\eta(t)$ as in the preceding problem, by pretending that G has reached its equilibrium value at time t. Show that in steady shearing starting at time zero, the stress is $\sigma(t) = G(t)\kappa(t) + \eta(t)\kappa'(t)$. Thus the stress is given at any time by a Voigt model, but the elastic modulus and viscosity coefficient that must be used are continually changing.

3. Solids and Fluids.

With steady shearing at the rate γ beginning at time zero, the stress at time t is

$$\sigma(t) = \gamma \int_0^t G(t')dt'.$$

If the integral converges in the limit as $t \to \infty$, then the stress approaches a steady-state value and we call the material a fluid. If the stress increases without bound, we call the material a solid.

To add some justification to these uses of the terms solid and fluid, let us consider a different kind of experiment. Suppose that a stress (possibly variable) is applied to the material from time zero to time t_1, and the material is then left stress-free. We will show that if the material is a solid in the previously-defined sense, then the material will return to its initial unsheared state after the load is removed, while if it is a fluid, there will be a permanent deformation $\kappa(\infty)$ generally not zero.

For times greater than t_1, after the load has been removed, the creep integral gives

$$\kappa(t) = \int_0^{t_1} J(t-t')d\sigma(t') = \int_0^{t_1} J'(t-t')\sigma(t')dt'.$$

Then, assuming that J' approaches a limiting value $J'(\infty)$, the shear approaches an equilibrium value

$$\kappa(\infty) = J'(\infty)\int_0^{t_1} \sigma(t)dt.$$

If $J'(\infty)$ is zero, recovery is complete, but if not, there is a residual deformation.

To establish the connection with the integrability of G, we note that the limit of J' is the value of its s-multiplied transform at $s = 0$. Hence, we obtain

$$J'(\infty) = \lim_{s\to 0} s(s\overline{J}-J_g) = \lim_{s\to 0} 1/\overline{G}(s).$$

If G is integrable from zero to infinity (fluid), the integral is $\overline{G}(0)$, and $J'(\infty) = 1/\overline{G}(0)$. Otherwise $\overline{G}(s)$ diverges as $s \to 0$, and $J'(\infty) = 0$.

Problem: For each of the following response functions, state whether the behavior is that of a solid or a fluid. Sketch the behavior of the other response function, using all relations at your disposal. (a) $G = t^{-\frac{1}{2}}$. (b) $G = t^{-\frac{1}{2}}\exp(-t)$. (c) $J = \log(t+1)$, (d) $J = 1 + t^{\frac{1}{2}} + t$.

4. Scale-Invariant Response.

It has been said that the relaxation time of a polymer is the time of observation, divided by π (Orowan's Rule). If we observe G from time zero to time t_o, we find that it decays sharply during about the first third of the interval and then remains fairly constant during the last two-thirds, no matter what the time of observation t_o may be, unless it is extremely long or short.

Let $t = xt_o$ so that x stands for the fraction of the total time of observation. Then as a function of x, $G(xt_o)$ has pretty much the same shape for all values of the parameter t_o. Similarly, the functions $J(xt_o)$ look like various multiples $M(t_o)$ of a single function $F(x)$. Let's pretend that this is literally true:

$$J(xt_o) = M(t_o)F(x).$$

We can solve this equation for J, M, and F. First, notice that by multiplying M and dividing F by the same constant, we can normalize M so that $M(1) = 1$, and it then follows that $F(x) = J(x)$:

$$J(ct) = M(c)J(t).$$

Then

$$M(ab) = \frac{J(abt)}{J(t)} = \frac{J(abt)}{J(bt)}\,\frac{J(bt)}{J(t)} = M(a)M(b).$$

If we write $\log M(e^x) = m(x)$, so that a plot of $m(x)$ is the same as $\log M$ versus $\log a$, we have

$$m(x+y) = m(x) + m(y).$$

This suggests that $m(x)$ is linear. In fact, if m is smooth then by differentiating with respect to y and then setting $y = 0$, we obtain $m'(x) = m'(0)$, so $m(x)$ has constant slope, p, say. From $M(1) = 1$ we find that $m(0) = 0$. Hence $m(x) = px$. Then $M(c) = c^p$; straight lines on a log-log plot give powers on an ordinary plot. Then $J(ct) = c^p J(t)$. Setting $t = 1$ and then calling c by the name of t instead, we obtain

$$J(t) = J(1)t^p.$$

Thus, powers are the only functions that are self-similar under scaling, or <u>scale-invariant</u>.

This has all been a very long way of saying that plots of $\log J$ or $\log G$ versus $\log t$ are usually fairly straight lines over a broad range of values of t. The slope, which represents the power, is usually small. The slope doesn't really remain constant for all values of t; the curves level off at the limiting values of $\log G$ or $\log J$ when $\log t$ goes to plus or minus infinity.

5. <u>Approximate Transform Inversion.</u>

We can use the power-like behavior of J and G to help us find approximate relations between them. This can be done directly, or with Laplace transforms as an intermediate step. We first consider a method based on a crude approximate transform inversion.

The s-multiplied transform is called a Carson transform. For real positive s, the Carson transform of J is

$$s\bar{J}(s) = \int_0^\infty \exp(-x)J(x/s)ds.$$

Evidently it is the average value of $J(x/s)$ with respect to the weight function $\exp(-x)$. (<u>Problem</u>: From the interpretation as an average, show that $s\bar{J}$ approaches J_g when s is large and J_e when s is small, if J_e exists.)

Assuming that J is like a small power t^p, it is fairly flat in the vicinity of $1/s$, so the average of $J(x/s)$ ought to be about equal to $J(1/s)$:

$$s\bar{J}(s) \cong J(1/s), \quad tJ(t) \cong \bar{J}(1/t).$$

We can do a bit better with hardly any more effort. For the average, take $J(x_o/s)$ where x_o is chosen so that the relation,

$$s\bar{J}(s) \cong J(x_o/s),$$

is accurate for small powers t^p. As it stands, it gives $p!/s^p = (x_o/s)^p$. The s-dependence cancels nicely, leaving $p! = x_o^p$. This is exact at $p = 0$ for any x_o. To make it accurate to first order in p, we equate the derivatives of both sides at $p = 0$. This gives $\log x_o = -\gamma$, where γ is Euler's constant, which is defined by

$$-\gamma = \left.\frac{d(p!)}{dp}\right|_{p=0} = \int_0^\infty \exp(-x)\log(x)dx.$$

Euler's constant is 0.5772 to four places, so $x_o = \exp(-\gamma) = 0.5615$. That looks reasonable. For the average value of $J(x/s)$ with respect to the weight $\exp(-x)$, we use the value at about $x = \frac{1}{2}$:

$$s\bar{J}(s) \cong J(0.56/s), \quad J(t) \cong (0.56/t)\bar{J}(0.56/t).$$

We will refer to this, or the same thing with $x_o = \frac{1}{2}$, as Schapery's Rule. (R. A. Schapery, Proc. 4th U.S. Nat'l Cong. of Appl. Mech. 2, 1075 (1962).)

6. Direct Approximation.

If we use the preceding method to evaluate the transforms of J and G, then from the reciprocal relation between the transforms we find that to the order of approximation considered, J and G are themselves reciprocal. This is a good rough idea, but it is easy to do better. For variety, let's work directly in terms of the integral equation instead of using transforms.

To determine G at time t, we need to know the values of J up to time t, $J(xt)$ $(0 \leq x \leq 1)$. Instead of picking a typical value on this interval as an approximation, as in the preceding method, let us approximate J by a power and try to pick the most appropriate power. To determine the power, we use a linear approximation to $\log J(xt)$ as a function of $\log x$:

$$\log J(xt) = \log J(t) + p(t) \log x.$$

Here $p(t)$ is the slope at t on a plot of $\log J$ versus $\log t$:

$$p(t) = \frac{d(\log J)}{d(\log t)} .$$

Then we have

$$J(xt) = J(t)x^{p(t)}.$$

Similarly, we suppose that G can be approximated for times up to t by

$$G(xt) = G(t)x^{-q(t)}.$$

Here, both $G(t)$ and the log-log slope $-q(t)$ are unknown.

The equation connecting J and G is

$$1 = \int_{-\infty}^{t} G(t-t')dJ(t') = \int_{0-}^{1} G[t(1-x)]dJ(tx),$$

where we have replaced t' by xt. By substituting the approximations for J and G, we obtain

$$\begin{aligned} 1 &= G(t)J(t) \int_{0}^{1} (1-x)^{-q} px^{p-1} dx \\ &= G(t)J(t)(-q)!p!/(p-q)!. \end{aligned}$$

We restrict attention to cases in which $p(t)$ does not vary rapidly, and hope that the same is true for $q(t)$. Then, treating the factorials as constants, we find that G is proportional to $1/J$. If so, then $q = p$. Hence,

$$J(t)G(t) = 1/p!(-p)! = \sin[p(t)\pi]/p(t)\pi.$$

If this "improved" approximation differs very much from the first approximation $JG \cong 1$, then neither one should be trusted. A first approximation that is worth anything at all cannot be improved very much.

For fluids at large times, p is close to unity and the approximation is very bad. For, if p is close to 1, then $-q$ is close to -1. Since $(-q)!$ is singular at -1, a small or slow variation of q will cause a large or fast variation of $(-q)!$, contrary to the assumption of slow variation used in obtaining the result.

Problem: Derive the preceding approximation by using transforms.

Example: Although this approximation is exact, for practical purposes, when p is small, it is certainly not universally valid. Consider a Maxwell fluid with $J = 1 + t$, for which the modulus is $G = \exp(-t)$. Simply taking G to be the reciprocal of J would give $G \cong 1/(1+t)$. Calculating $p(t)$ is more informative. We obtain $p(t) = t/(1+t)$. Since p is small when t is small, both $G \cong 1/J$ and the improved approximation should be good when t is small. For large t, p comes close to unity, so both approximations will be bad. The "improved" approximation is

$$G(t) \cong \frac{1}{\pi t} \sin \frac{\pi t}{1+t} .$$

At $t = 0$, where p is zero, this starts off with the right value and the right slope, and so does $1/J$. At $t = 1$ (where $p = \frac{1}{2}$) the exact and approximate values are 0.37 and 0.32; $1/J$ is 0.50. In other words, the "improved" approximation is better, but not very good. For large t, the approximation is asymptotic to $1/t^2$, so although the absolute error is small, the relative error becomes enormous as p approaches unity.

Problems.

1. Working the approximation the other way, start with $G(t) = \exp(-t)$. Show that the log-log slope $-q$ becomes unity at $t = 1$. Compare $J \cong 1/G$ and the second approximation with the exact compliance, for various ranges of values of t.

2. For a Kelvin-Voigt model $J = 1 - \exp(-t)$, compute $p(t)$. State where you think the approximation will be worst, giving your reasons. Compare the exact and approximate evaluations of $G(t)$.

3. A modulus of the form $G(t) = 1 + 9\exp(-t)$ is the least power-like that you will ever see in real life, among solids. Compute the log-log slope $-q$, and find out where it is largest. Compare $1/G$ and the second approximation with the exact values of J, and make up a moral to this story.

4. The following problem is not important for moduli and compliances, but it is important for some other functions that are related in the same way that the modulus is related to the compliance. Allow the un-physical possibility that J_g might be zero. Assume that $J(ct)/J(t)$ approaches a limit $M(c)$ when t approaches zero, for all positive values of c. Prove that $M(c) = c^p$, for some power p. Recall the expression for $s\bar{J}(s)$ as an average value of $J(x/s)$. Prove that when $s \to \infty$, $s\bar{J}$ is asymptotic to $p!J(1/s)$ if $p > -1$. What does $p > -1$ have to do with it? Similarly, if $G(ct)/G(t)$ approaches c^{-q} when $t \to 0$, then $s\bar{G}(s)$ is asymptotic to $(-q)!G(1/s)$ when $s \to \infty$ if $q < 1$. Thus, show that our numerical approximation formula is a precise asymptotic relation when $t \to 0$. Repeat this whole development for $t \to \infty$ and $s \to 0$. Conclude that if J is scale-invariant, so that no time is either "small" or "large" by any internal standard, then the asymptotic results must be true at all times.

7. Approximate Relation Between Modulus and Complex Modulus.

We can also obtain a simple approximate relation between $G(t)$ and $G^*(\omega)$ by assuming that G is power-like. In the expression for $s\overline{G}$ we set $s = re^{i\theta}$, and make the change of variable $rt = x$:

$$s\overline{G}(s) = e^{i\theta} \int_0^\infty \exp(-x \cos \theta - ix \sin \theta) G(x/r) dx.$$

At $s = i\omega$, r is equal to ω and $\theta = \pi/2$. As θ changes from 0 to $\pi/2$, the exponential changes from a pure decay $\exp(-x)$ to a pure oscillation $\exp(-ix)$. For $\theta = 0$ the values of G near $x = 1$ are the most relevant because values at larger x are damped out by the factor $\exp(-x)$. For $\theta = \pi/2$ the values near $x = 1$ are still the most important because the oscillation causes the values at larger x to cancel one another out. Consequently, let us approximate G by a power that fits well near $x = 1$:

$$G(x/r) \cong G(1/r) x^{-p}.$$

Here $-p$ is the log-log slope at $t = 1/r = 1/\omega$. Then,

$$s\overline{G}(s) \cong G(1/r)(-p)!(s/r)^p, \qquad r = |s|.$$

By evaluating this at $s = i\omega$, we obtain

$$G^*(\omega) \cong G(1/\omega)(-p)! \exp(ip\pi/2).$$

Then the storage modulus is

$$G_1(\omega) \cong G(1/\omega)(-p)! \cos \delta(\omega),$$

where the loss angle δ is

$$\delta(\omega) \cong p\pi/2 = -\frac{\pi}{2} \frac{d(\log G)}{d(\log t)}\bigg|_{t=1/\omega}.$$

Since G_1 and G are so directly related, the loss angle is also given by

$$\delta(\omega) \cong p\pi/2 \cong \frac{\pi}{2} \frac{d(\log G_1)}{d(\log \omega)} .$$

8. Graphs of Moduli and Compliances.

Because J^* and G^* are exactly reciprocal, then if the loss angle is small, J_1 and G_1 are approximately reciprocal. The loss angle is small if J and G are approximated well by small powers of t. In that case, J and G are also approximately reciprocal. Furthermore, G and G^* are roughly equal at reciprocal values of their arguments.

Taking the reciprocal of a quantity corresponds to changing the sign of its logarithm. Consequently, the reciprocal relations mentioned above mean that on a log-log plot, the curves of $\log J$ versus $\log t$, $-\log G$ versus $\log t$, $\log J_1$ versus $-\log \omega$, and $-\log G_1$ versus $-\log \omega$ are all roughly the same curve. This is something of an exaggeration, but it is nearly enough true to be a helpful idea. The loss angle is roughly $\pi/2$ times the slope of this curve.

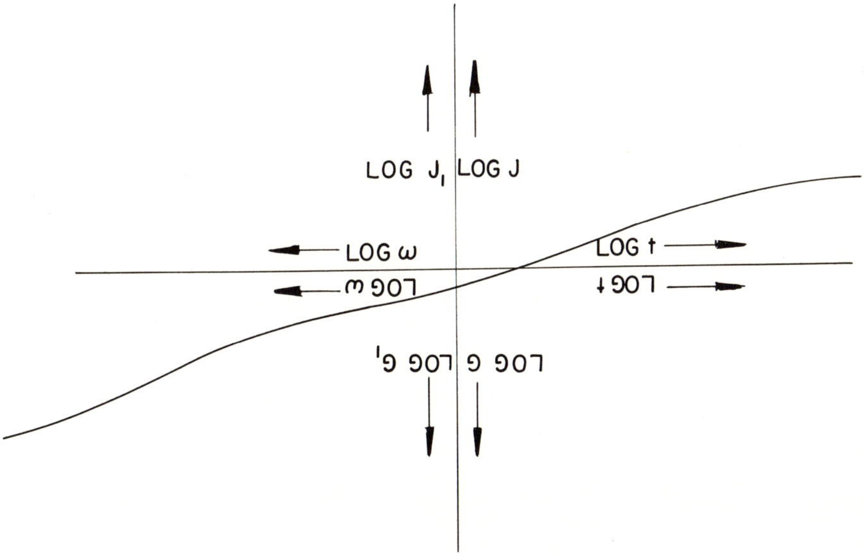

CHAPTER IV

SOME ONE-DIMENSIONAL DYNAMICAL PROBLEMS

We now consider some dynamical problems involving especially simple geometries: torsional oscillations of a rod, and propagation of plane waves in a semi-infinite medium. We have several purposes in mind. First, these problems illustrate some important similarities between the behavior of viscoelastic materials and the behaviors of materials that are purely viscous or purely elastic, and they also illustrate some important distinctions. Second, we wish to illustrate that viscoelasticity problems can be analyzed to significant depth even if the basic response functions are given only graphically or numerically, in the form of data. Third, the mathematical techniques used in these problems are of interest in themselves because they find applications in many areas.

Torsional oscillations are considered rather briefly. The greater part of this chapter is spent on the problem of one-dimensional pulse propagation and the various mathematical methods that can be illustrated in this context.

1. Torsional Oscillations.

We consider a rod of circular cross-section with radius R and length L. One end of the rod is attached to a rigid support. A fly-wheel with moment of inertia I is attached to the other end. At time zero the fly-wheel is turned through an angle θ_0 and then released. We expect that its angular displacement $\theta(t)$ will then be a damped sinusoidal oscillation. Given the modulus $G(t)$ of the material of the rod, we wish to determine the frequency and damping coefficient for the oscillation. Conversely, we might wish to determine the material properties by observing the oscillation.

We neglect the inertia of the rod itself and ignore gravity. With neglect of inertia we can suppose that the rod is twisted uniformly through the angle $\theta(t)/L$ per unit length at any given time. (Would you be willing to believe this if the rod were very short? A mile long? What sort of phenomenon is being neglected?)

Each material element (bounded by cylindrical coordinate surfaces) is sheared in the azimuthal direction when the rod is twisted. At a distance r from the axis, the amount of shear is $\kappa(r,t) = r\theta(t)/L$. (Show why, geometrically.) The shearing stress on a cross-section is then also azimuthal, with a resultant couple $M(t)$, say, which is related to $\theta(t)$ by

$$M(t) = (\pi R^4/2L) \int_{-\infty}^{t} G(t-t')d\theta(t').$$

(Problem: Derive this relation.)

The moment exerted on the fly-wheel by the rod is $-M(t)$, so the equation of motion of the fly-wheel is

$$I\theta''(t) = -M(t).$$

As initial conditions we suppose that θ is zero for all negative times, and that at $t = 0+$, $\theta = \theta_0$ and $\theta' = 0$. Then, for $t > 0$, we have

$$\theta''(t) + F\theta_0 G(t) + F \int_0^t G(t-t')\theta'(t')dt' = 0,$$

where $F = \pi R^4/2LI$.

By applying the Laplace transform and rearranging, we obtain

$$\frac{\bar{\theta}(s)}{\theta_0} = \frac{s}{s^2 + sF\bar{G}(s)} .$$

Our problem is to invert this transform. Before considering the general case, let's look at some examples.

a. Viscous fluid.

For purely viscous response there should be no oscillation at all. With $G(t) = \eta_0\delta(t)$, then $\bar{G}(s) = \eta_0$, and inverting the transform yields $\theta(t)/\theta_0 = \exp(-F\eta_0 t)$. This is obviously wrong. What is wrong with the derivation of it?

b. Elastic solid.

If $G(t) = G_e$, constant, then with $\omega^2 = FG_e$ we find that the transform is

$$\frac{\bar{\theta}(s)}{\theta_o} = \frac{s}{s^2 + \omega^2} = \frac{1}{2}\,\frac{1}{s + i\omega} + \frac{1}{2}\,\frac{1}{s - i\omega}\,.$$

If we invert the transform by using the inversion integral, then the poles at $s = \pm i\omega$ contribute the residues $\frac{1}{2}\exp(\pm i\omega t)$ to the inverse, and we obtain $\theta(t) = \theta_o \cos(\omega t)$. For the elastic material there is an undamped oscillation, as expected, and even the frequency could have been obtained by dimensional analysis. In more complicated cases we will look for poles of the transform, such as those found here, in order to determine the frequency of oscillation.

c. Maxwell fluid.

As the simplest case that includes some damping, we consider the modulus $G(t) = G_g \exp(-t/T)$. From the parameters of the problems, R, L, I, θ_o, G_g, and T, we can form three dimensionless parameters; θ_o itself, the slenderness R/L, and a combination FG_gT^2. If we define $\omega_o^2 = FG_g$, the last of these parameters can be replaced by $\omega_o T$, a frequency of oscillation made dimensionless with respect to a time scale intrinsic to the material. Now, intuitively, if T is very large then the stress does not relax much in any reasonable amount of time, so the material should behave as if it were elastic, with modulus G_g. Of course, since T is dimensional it is meaningless to talk about "T large" unless we say what it is large in comparison to. The dimensional analysis tells us that what we should really consider is large values of $\omega_o T$. When this parameter is large, we should expect oscillation at a frequency near ω_o. At the opposite extreme, if $\omega_o T$ is small there will probably be overdamping, pure decay with no oscillation at all.

Problem: Show that if $\omega_o T$ is large, $\bar{\theta}(s)$ has poles near $-(1/2T) \pm i\omega_o$. Show that $\theta(t)$ is then a damped oscillation with frequency near ω_o and amplitude approximately $\theta_o \exp(-t/2T)$. Show that if $\omega_o T$ is small, then the transform has poles on the negative real axis, near $-1/T$ and $-(\omega_o T)^2/T$. Show that the former contributes nothing of any significance to $\theta(t)$. Show that $\theta(t)$ is nearly constant over times short in comparison to $T/(\omega_o T)^2$, but ultimately decays to zero.

d. Power-law solid.

With $G(t) = Ct^{-p}/(-p)!$, then $s\bar{G}(s) = Cs^p$. The inversion integral for $\theta(t)$ is

$$\frac{\theta(t)}{\theta_o} = \frac{1}{2\pi i} \int \frac{se^{st}ds}{s^2 + FCs^p} .$$

The function s^p, and thus the integrand, has a branch cut along the negative real axis. In the cut plane, the integrand has poles at $s = -r \sin \alpha \pm ir \cos \alpha$, where

$$r^{2-p} = FC \quad \text{and} \quad \alpha = \frac{\pi}{2} \frac{\frac{1}{2}p}{1 - \frac{1}{2}p} .$$

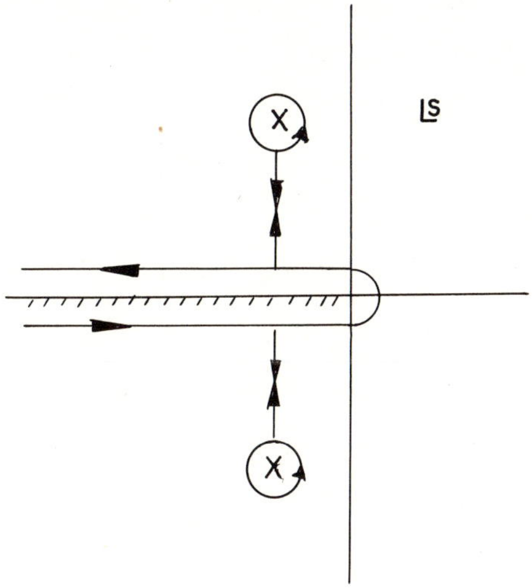

The inversion contour can be deformed into a loop integral along the two sides of the branch cut, plus circles around the two poles. The poles contribute a damped oscillation at the frequency $\omega = r \cos \alpha$, decaying as $\exp[-(r \sin \alpha)t]$. The integrals along the two sides of the branch cut can be combined into a single real integral,

$$-\frac{FC \sin(p\pi)}{\pi} \int_0^\infty \frac{x^{1+p}\exp(-xt)dx}{x^4 + 2FCx^{2+p}\cos(p\pi) + (FC)^2x^{2p}} .$$

In spite of the complicated appearance of this integral, it is evident that as a function of t, it is a linear combination of pure decays $\exp(-xt)$ with various values of x. If p is small, it is roughly the same as

$$-FCp \int_0^\infty \frac{x \exp(-xt)}{(x^2 + FC)^2} dx,$$

and thus it is only a small contribution proportional to p. Also, the oscillation is only lightly damped when p is small. Consequently, the pure decay contributed by the loop integral can be neglected in comparison to the damped oscillation contributed by the poles, provided that p is small, i.e., the material is nearly elastic. In that limit, the frequency is $\omega \cong (FC)^{\frac{1}{2}}$ and the damping constant is $\omega(p\pi/4)$. In one cycle of oscillation, the logarithm of the amplitude decreases by $p\pi^2/2 \cong 5p$, independent of ω.

e. Real solids with small losses.

For solid polymers we can model the behavior of the modulus in the vicinity of a time t_o by a power law, $G(t) = G(t_o)(t/t_o)^{-p(t_o)}$. We can guess that in the present problem, the most appropriate time t_o to use in this approximation must be something of the order of the period of oscillation, but we don't really know exactly, so it is not entirely obvious what values of C and p we should use in the preceding results in order to get the right answer for a given real material.

The frequency and decay constant are to be determined as the imaginary and real parts of the roots of $s^2 + Fs\overline{G}(s) = 0$. We suppose that the material is not too lossy, so that these roots lie close to the imaginary axis. Let us write $s = i\omega \exp(i\alpha)$ for the root near $i\omega$; the other is the complex conjugate of this. Then, if we also express $s\overline{G}$ in polar form as $|s\overline{G}|\exp(i\delta)$, we obtain

$$\omega^2 = |s\overline{G}|F \quad \text{and} \quad \alpha = \tfrac{1}{2}\delta.$$

We solve these equations by iteration, taking $\alpha = 0$ as a first approximation. With $s = i\omega$, $s\overline{G}$ is just the complex modulus $G^*(\omega)$. Then the first equation above is

$$\log|G^*(\omega)| = 2 \log \omega - \log F.$$

Given a graph of $\log|G^*|$ versus $\log \omega$, (or G_1, since $G_1 \cong |G^*|$), the frequency ω is easily determined as the intersection of this curve with the straight line $y = 2 \log \omega - \log F$. The loss angle δ at this frequency is then approximately $\frac{1}{2} p\pi$, where p is the slope of the log-log plot at the frequency found, and thus $\alpha = p\pi/4$ is obtained as a second approximation.

Problem: Expand the analytic function $\log(s\overline{G})$ about the point $s = i\omega$ in powers of $\log(s/i\omega)$, up to terms of second order. Express all coefficients in terms of the complex modulus and its log-log slope. What has been neglected in the method of solution outlined above? Where, if anywhere, was the assumption of power-like behavior used?

2. Plane Shear Waves.

We now turn to a problem in which the inertia of the viscoelastic material itself must be considered. We suppose that a semi-infinite body of material occupies the region $y \geq 0$. The boundary $y = 0$ is moved tangentially to itself, in the x-direction, with velocity $v_o(t)$. This causes a disturbance which propagates into the body, so that particles move in the x-direction with velocity $v(y,t)$. Since all equations to be used are linear, we can express $v(y,t)$ in terms of $v_o(t)$ by a linear input-output relation,

$$v(y,t) = \int_{-\infty}^{t} V(y,t-t')v_o(t')dt'.$$

The response function $V(y,t)$ in this relation is evidently the velocity induced by a pulse in wall velocity, $v_o(t) = \delta(t)$, i.e., a unit step in wall displacement. Our problem is to find $V(y,t)$, the pulse response function.

Problem: Regard $V(y,t)$ as well as $J(t)$ and $G(t)$ as known. Express (a) the stress $\sigma(y,t)$ in terms of the wall velocity, (b) the shear rate $v_y(y,t)$ in terms of the wall displacement $u_o(t)$, and (c) the stress in terms of the wall stress. It is convenient to use the convolution notation.

Shearing the material will cause heating. Since viscoelastic material properties are temperature-dependent, a more realistic treatment would require study of the coupled thermo-mechanical problem. Heating also causes thermal expansion, which in turn creates a compression wave in addition to the shear wave. Here we treat material properties as if they were temperature-independent, and we ignore the compression wave.

Since the rate of shear is $\kappa_t(y,t) = v_y(y,t)$, then the shearing stress is

$$\sigma(y,t) = \int_{-\infty}^{t} G(t-t')v_y(y,t')dt'.$$

We use this in the x-component of the momentum equation, which is

$$\rho v_t(y,t) = \sigma_y(y,t).$$

Here ρ is the mass density.

Assuming that the medium is quiet at time zero, then by applying the Laplace transform we obtain

$$\rho s\overline{v}(y,s) = \overline{G}(s)\overline{v}_{yy}(y,s).$$

The boundary conditions are $\overline{v}(0,s) = \overline{v}_o(s)$ and $\overline{v}(\infty,s) = 0$. Then, with $\overline{J} = 1/s^2\overline{G}$, we obtain

$$\overline{v}(y,s) = \overline{v}_o(s)\exp\left(-ys\sqrt{\rho s\overline{J}}\right).$$

By applying the convolution theorem we now obtain the input-output relation mentioned earlier. The kernel $V(y,t)$ in it, which is the response to a Dirac delta in wall velocity, is identified as the function whose transform is

$$\overline{V}(y,s) = \exp\left(-ys\sqrt{\rho s\overline{J}}\right).$$

Our problem now is to invert this transform. We will look at some special cases before considering the general problem.

Problem: A semi-infinite rod of radius R occupies the region $r \leq R$, $z \geq 0$. The end $z = 0$ is subjected to an angular displacement $\theta_o(t)$, causing a generic cross-section to turn through the angle $\theta(z,t)$. (a) Express the moment $M(z,t)$ on a generic cross-section in terms of $G(t)$, $\theta(z,t)$, and geometrical factors. (b) Find the moment of inertia of the slice of material between z and $z + dz$. (c) Set up the (angular) momentum equation. (d) Reduce the problem to that of inverting a transform.

a. Elastic solid.

With $J(t) = J_e$, then $s\bar{J}(s) = J_e$ as well, and $\bar{V}$ is $\exp(-sy/c_e)$, where $c_e = (\rho J_e)^{-\frac{1}{2}} = (G_e/\rho)^{\frac{1}{2}}$. Thus, $V = \delta(t-yc_e^{-1})$. The pulse travels into the medium at a constant speed c_e, with no dispersion or attenuation. A general wall motion $v_o(t)$ yields a motion $v(y,t) = v_o(t-yc_e^{-1})$ in the medium.

b. Viscous fluid.

With $J = t/\eta$, then $s\bar{J} = 1/\eta s$, so $\bar{V} = \exp(-ys^{\frac{1}{2}}/\nu^{\frac{1}{2}})$, where $\nu = \eta/\rho$ is the kinematic viscosity. Let us invert $\exp(-s^{\frac{1}{2}})$. We can then recover V by using the scaling formula. The inversion integral is

$$\frac{1}{2\pi i} \int \exp(st-s^{\frac{1}{2}})ds.$$

Let's evaluate this by using a saddle-point approximation. We first locate the place s_o where the exponent is stationary: $d(st-s^{\frac{1}{2}})/ds = 0$. This gives $s_o = 1/4t^2$ as the only saddle point. The second derivative there is $(\frac{1}{2})^2 s_o^{-3/2} = 2t^3$. Then the exponent is given near the saddle point by $-(1/4t) + t^3(s-s_o)^2 + \cdots$. If we scale the distance from the saddle point by introducing a new variable $z = t^{3/2}(s-s_o)$, then the integral becomes

$$\frac{1}{2\pi i} \frac{\exp(-1/4t)}{t^{3/2}} \int \exp(z^2 + \cdots)dz.$$

Evidently the integrand grows very large on both sides of the saddle point as z varies through real values. As z varies in the pure imaginary direction, $z = iy$, the integrand grows small very quickly, like $\exp(-y^2)$. This is why we call the point $z = 0$ a saddle point.

We can shift the inversion contour over until it passes through the saddle point. We want the integrand to be negligible except just when climbing up and over the col, so the path should be in the pure imaginary direction, but it already is. Putting $z = iy$, we get

$$\int \exp(z^2)dz = i \int_{-\infty}^{\infty} \exp(-y^2)dy = i\sqrt{\pi}.$$

Thus, $\exp(-s^{\frac{1}{2}})$ is the transform of

$$L^{-1}[\exp(-s^{\frac{1}{2}})] = \frac{1}{2\sqrt{\pi}} \frac{\exp(-1/4t)}{t^{3/2}},$$

at least approximately.

<u>Problem</u>: Show that this result is exact. In the original integral, set $z = t^{\frac{1}{2}}(s^{\frac{1}{2}}-s_o^{\frac{1}{2}})$. Worry about the integration contour.

By using the scaling formula, $\bar{f}(s/c) = L[cf(ct)]$, we obtain

$$V(y,t) = \frac{Y\exp(-Y^2/4)}{2\sqrt{\pi}t}, \text{ where } Y = y/(\nu t)^{\frac{1}{2}}.$$

The pulse has the same shape at all times but spreads out in proportion to $t^{\frac{1}{2}}$ and decays in proportion to $1/t$. At time zero the pulse width is zero and the amplitude is infinite. This is the Dirac delta in wall velocity. Later, as the pulse spreads, the maximum velocity occurs at the point $y = (2\nu t)^{\frac{1}{2}}$. The pulse does not really detach from the wall. The signal <u>diffuses</u> into the medium rather than translating as a well-defined wave.

c. Power-law solids.

We can consider a range of behavior from purely elastic $(p = 0)$ to purely viscous $(p = 1)$ by considering compliances of the form $J(t) = t^p/Cp!$. Since the corresponding moduli $G(t) = Ct^{-p}/(-p)!$ are infinite at time zero (unless $p = 0$), then according to the model it requires infinite stress to produce a step change in strain, independent of any consideration of inertia. The instantaneous response is more like that of a viscous fluid than that of an elastic solid. Consequently, a pulse in wall velocity will be felt immediately at all distances from the wall, as in the preceding example. Although this is unrealistic, the model may nevertheless give us a good idea of what happens to the main part of the signal, where most of the momentum is. Later we will come back to the question of instantaneous response.

With $s\bar{J}$ equal to $1/Cs^p$, the transform to be inverted is

$$\bar{V}(y,s) = \exp[-ys^{1-\frac{1}{2}p}(\rho/C)^{\frac{1}{2}}].$$

I doubt that this can be inverted in terms of known functions; certainly I do not know how to do so. However, the more important qualitative and quantitative features of the pulse can be deduced from the transform without inverting it explicitly.

Let us write $K(t;p)$ for the function whose transform is $\exp(-s^{1-\frac{1}{2}p})$. This function represents the time history of the signal received at the station $y_1 = (C/\rho)^{\frac{1}{2}}$. We have already found the form of this function in two special cases:

$$K(t;0) = \delta(t-1) \qquad \text{(elastic solid)},$$
$$K(t;1) = \left(2\sqrt{\pi}\right)^{-1} t^{-3/2}\exp(-1/4t) \qquad \text{(viscous fluid)}.$$

The history of the signal at an arbitrary station y is found by using the scaling formula:

$$V(y,t) = (y_1/y)^{1+\beta}K[t(y_1/y)^{1+\beta};p].$$

Here $1 + \beta$ is the reciprocal of $1 - \frac{1}{2}p$. We can immediately draw the interesting conclusion that the pulse shape (in time) is the same at all stations. The pulse width varies from one station to another in proportion to $y^{1+\beta}$; it is short at stations close to the source and longer at stations farther away. The amplitude varies in inverse proportion to the width, so the area under a plot of V versus t is the same at every station. (Problem: What is this area?)

Kolsky (Philosophical Magazine, August, 1956) discovered this phenomenon of a universal pulse shape in experiments on polyethylene, polystyrene, and polymethylmethacrylate, and he explained it theoretically as a consequence of the near-constancy of the loss angle over a broad range of frequencies. In our terms, this means that the compliance is approximately a power t^p over a broad range of time. Kolsky also found that the pulse shape is not only the same at different stations in the same material, but is also the same for different materials (different values of p) after an appropriate scaling, at least if p is small. (Kolsky's experiments involved longitudinal stress pulses in rods, but the mathematics is all the same as for shear wave propagation.)

The general form of the result also shows how the pulse speed varies during the motion. Let t_m (presumably depending on p) be the value of t that maximizes $K(t;p)$. Then the signal has maximum amplitude at station y when $t(y_1/y)^{1+\beta} = t_m$. If we define a signal speed by the speed of this locus, then after a little manipulation we find that

$$dy/dt = [\rho J(t)]^{-\frac{1}{2}} n(p),$$

where $n(p)$ is some pure number depending on p. For the two cases in which we already know the pulse shape explicitly, we find that $n(0) = 1$ and $n(1) = \frac{1}{2}\sqrt{6} = 1.2$. (Problem: Verify this.) Other definitions of the pulse speed, such as the speed of the locus of the maximum with respect to y, give results different from this one only in the numerical factor. The main result is that the pulse speed is roughly $[\rho J(t)]^{-\frac{1}{2}}$, so the pulse slows down as time progresses. If p is small, so that J appears to be nearly constant from $t_o/10$ to t_o, say, where t_o is the total time of observation, then an experiment will produce the appearance that the pulse velocity is constant. Which constant depends on the time of observation, however.

By multiplying and dividing the expression for V by t, and then defining a new function F whose argument is an appropriate negative power of that of K, we can put the result into the form

$$V(y,t) = t^{-1}F(Y;p),$$

where Y is a scaled y-coordinate:

$$Y = (y/t^{1-\frac{1}{2}p})(\rho/C)^{\frac{1}{2}}.$$

The function F is given in the two extreme cases by

$$F(Y;0) = \delta(Y-1) \qquad \text{(elastic solid)}$$

and

$$F(Y;1) = \left(2\sqrt{\pi}\right)^{-1} Y \exp(-Y^2/4) \qquad \text{(viscous fluid)}.$$

This form of the result makes it clear that the spatial shape of the pulse is the same at each instant, the only changes being due to time-dependent changes of width and amplitude. A feature that was not previously so apparent comes out very clearly here: <u>The maximum amplitude (in space) varies precisely as $1/t$</u> for all p and thus, by exaggeration, for all real materials.

To obtain more quantitative information about the pulse shape, let us consider approximations valid at short and long times. First, from the transform $\exp(-s^{1-\frac{1}{2}p})$ we conclude that $K(t;p)$ and all of its derivatives are zero at $t = 0$. (<u>Problem</u>: Prove this.) Thus K starts off very flat, like $\exp(-1/t)$, say. We will return to the short-time behavior later.

Turning to the long-time behavior, we consider the behavior of the s-multiplied transform for s approaching zero. This vanishes, so $K(\infty;p) = 0$, which is hardly surprising; the signal dies out after a long time. Indeed, since the transform itself is equal to unity at $s = 0$, then K dies out fast enough to be integrable:

$$\int_0^\infty K(t;p)dt = 1.$$

Thus it dies out faster than $1/t$, in particular. Now, the transform of $tK(t;p)$ is $(1-\frac{1}{2}p)s^{-\frac{1}{2}p}\exp(-s^{1-\frac{1}{2}p})$. For s approaching zero this is asymptotic to $(1-\frac{1}{2}p)s^{-\frac{1}{2}p}$, which is the transform of $t^{-1+\frac{1}{2}p}(1-\frac{1}{2}p)/(\frac{1}{2}p-1)!$ Hence,

$$K(t;p) \sim t^{\frac{1}{2}p-2}\tfrac{1}{2}p(1-\tfrac{1}{2}p)/(\tfrac{1}{2}p)! \qquad (t \to \infty).$$

<u>Problem</u>: Let K_1 denote this approximation. Find the transform of $t^2(K-K_1)$. From its behavior as $s \to 0$, find the asymptotic behavior of $t^2(K-K_1)$ for large t. Thus obtain the second term of an asymptotic expansion of K. Can you extend the method to obtain further terms?

1. <u>Saddle-Point Approximation.</u> We can try to get a more detailed picture of the function $K(t;p)$ by performing an approximate inversion of the transform $\exp(-s^{1-\epsilon})(\epsilon = \frac{1}{2}p)$, using the saddle-point method. However, it turns out that the saddle point approximation is valid only for small t, as we shall see.

The inversion integral is

$$K(t;p) = (2\pi i)^{-1} \int \exp(st-s^{1-\epsilon})ds.$$

The exponent is stationary at the point s_o where $t = (1-\epsilon)s_o^{-\epsilon}$. The value of the exponent there is

$$-Z_o = -\epsilon s_o^{1-\epsilon} = -\epsilon[(1-\epsilon)/t]^{(1/\epsilon)-1}.$$

By expanding the exponent in powers of $s - s_o$ we obtain

$$-Z_o + \tfrac{1}{2}\epsilon(t/s_o)(s-s_o)^2 - (1/6)\epsilon(1+\epsilon)(t/s_o^2)(s-s_o)^3 - \cdots .$$

The change of variable $z^2 = \epsilon(t/s_o)(s-s_o)^2$ reduces this to

$$-Z_o + \tfrac{1}{2}z^2 - (1/6)(1+\epsilon)(1-\epsilon)^{-\frac{1}{2}}Z_o^{-\frac{1}{2}}z^3 - \dots .$$

We have kept the third-order term, temporarily, in order to determine when the saddle-point approximation will be valid. We see that Z_o should be large in order to neglect the cubic term. Thus, t should be small. The resulting approximation is accordingly valid asymptotically as t approaches zero:

$$K(t;p) \sim (2\pi)^{-\frac{1}{2}}(\epsilon t)^{-1}(1-\epsilon)^{\frac{1}{2}}Z_o^{\frac{1}{2}} \exp(-Z_o).$$

Problems.

1. Fill in the details.

2. Verify that this result is exact for all t if $p = 1$ (viscous fluid).

3. Show that if $p \neq 1$, the present small-t approximation is not the same as the large-t approximation obtained earlier.

4. Show that when ϵ is small, Z_o is approximately $(\epsilon/e)t^{1-(1/\epsilon)}$.

5. Draw a graph of the function on the right-hand side of the equation above, for $p = 0.2$, for all t and not merely for t small. How large do you suppose t can be before the approximation goes bad?

2. Wide Saddle-Points. When p (2ϵ) is small, the approximation that we have obtained depends on t much more strongly through Z_o than through the explicit factor $1/t$. Ignoring the latter, we find that the pulse has maximum amplitude at $Z_o = \frac{1}{2}$. This is probably too small a value of Z_o for the saddle-point approximation to be very accurate, so it is not likely that our result gives the pulse shape correctly.

Kolsky's experiments, mentioned previously, involved materials with very small values of p. The case of small p is mathematically interesting because at $p = 0$ the pulse is a Dirac delta, so a small-p approximation is a singular perturbation. We want to scale the pulse with respect to p in such a way that for small p

we can see some structure, and not just a Dirac delta.

In the saddle-point method we expanded the exponent in powers of $s - s_0$ and neglected terms of third degree and higher. If Z_0 is small this is not accurate. Let's try to get a more accurate approximation for the behavior of the exponent near the saddle point, by supposing that ϵ is small rather than that Z_0 is large.

We write the exponent as

$$st - s^{1-\epsilon} = st - s(s_0^{-\epsilon})(s/s_0)^{-\epsilon}.$$

The factor $s_0^{-\epsilon}$ is nearly independent of ϵ, because the saddle-point s_0 is defined by $s_0^{-\epsilon} = t/(1-\epsilon)$. Consequently, we leave this factor alone. For the factor $(s/s_0)^{-\epsilon}$, however, with s near s_0, we use the approximation

$$(s/s_0)^{-\epsilon} = 1 - \epsilon \log(s/s_0) + O(\epsilon^2).$$

Then the exponent is approximately

$$st - s^{1-\epsilon} = -\epsilon s_0^{-\epsilon} s[1 - \log(s/s_0)].$$

Then, defining a new integration variable $Z = \epsilon s_0^{-\epsilon} s$, and noticing that Z_0 is the value of this at $s = s_0$, we can write the exponent as $-Z[1 - \log(Z/Z_0)] = Z \log(Z/Z_0 e)$. Then, for ϵ small we have

$$K(t;p) \sim (\epsilon t)^{-1} I(Z_0) \qquad (\epsilon \to 0),$$

where $I(Z_0)$ is the function defined by the integral

$$I(Z_0) = (2\pi i)^{-1} \int \exp[Z \log(Z/Z_0 e)] dZ.$$

Comparison with the saddle-point approximation shows that when Z_0 is large, the function $I(Z_0)$ is approximately

$$I(Z_0) \sim (2\pi)^{-\frac{1}{2}} Z_0^{\frac{1}{2}} \exp(-Z_0).$$

We still do not know what the function $I(Z_0)$ is like for small values of Z_0. Nevertheless, we have accomplished something by expressing $K(t;p)$, a function of two variables, in terms of $I(Z_0)$, a function of one variable, in which we know explicitly how Z_0 depends on t and p (or ϵ). Thus, for small ϵ we have

$$K(t;p) \sim (\epsilon t)^{-1} I[(\epsilon/e)t^{1-(1/\epsilon)}].$$

The argument of I varies with enormous rapidity as t varies, so I goes through practically all of its values within some small neighborhood of the time t_m at which it is maximum. If the maximum of I occurs at $Z_0 = Z_m$, (a definite number independent of ϵ), then

$$t_m = (\epsilon/eZ_m)^{\epsilon/(1-\epsilon)} \sim \epsilon^{\epsilon}.$$

Thus, the time of arrival of the signal peak is $\exp(\epsilon \log \epsilon)$ to lowest order, and it turns out that we don't need to know what Z_m is. As a check, we verify that at $\epsilon = 0$ the peak time is $t_m = 1$. However, for ϵ small it is $t_m = 1 - \epsilon \log(1/\epsilon)$, so $t_m = 1$ is not a very good approximation unless ϵ is extremely small. In other words, even if p is small, viscous effects make the pulse go noticeably faster than if the material were perfectly elastic.

We can draw another conclusion about the pulse shape, still without knowing the form of the function $I(Z_0)$ exactly. Near the peak time $t = t_m$, the argument of I is

$$Z_o = Z_m - \frac{Z_m}{t_m} \frac{t-t_m}{\epsilon} + \cdots$$

Since ϵt_m is just ϵ to lowest order, it follows that in the neighborhood of its peak, the pulse shape is a function of $(t-t_m)/\epsilon$. Thus, the width of the peak is proportional to ϵ. Since $\epsilon = \frac{1}{2}p = \delta/\pi$ where δ is the loss angle, this agrees with Kolsky's conclusion that pulse shapes in different materials are the same if brought to the same width with the loss tangent as a scale factor. From the present results we expect that the low-amplitude parts of the signals will not agree under this scaling, but of course this is not especially important.

d. Real solids with small losses.

In discussing the power-law model we have mentioned experimental results. The power-law model agrees with data for hard polymers quite well, but the power p that fits best generally varies slowly with time or frequency, so, for a given real material it is not entirely obvious which value of p (and C) should be used in the preceding results. We now consider a method that avoids explicit mention of power-law behavior; the appropriate power sneaks in sideways.

Before discussing pulse propagation, let us consider a different form of wave motion, the motion produced by a sinusoidally oscillating boundary, $v_o(t) = A \exp(i\omega t)$ (real part). If we go back to the original equations of motion and try a sinusoidal (exponential) solution, we find that the equations and boundary condition are satisfied by

$$v(y,t) = A \exp[i\omega(t-c^{-1}y) - \alpha y],$$

provided that the attenuation constant $\alpha(\omega)$ and the phase velocity $c(\omega)$ are given by

$$\omega c^{-1} - i\alpha = \omega(\rho J^*)^{\frac{1}{2}} = \omega(\rho|J^*|)^{\frac{1}{2}}(\cos\tfrac{1}{2}\delta - i\sin\tfrac{1}{2}\delta).$$

If the loss angle $\delta(\omega)$ is small then the phase velocity is determined approximately by the storage compliance or storage modulus:

$$c(\omega) \cong [\rho J_1(\omega)]^{-\frac{1}{2}} \cong [G_1(\omega)/\rho]^{\frac{1}{2}}.$$

Whether the loss angle is small or not, the attenuation constant can be written as

$$\alpha(\omega) = \omega c^{-1}(\omega) \tan\tfrac{1}{2}\delta(\omega).$$

Since the storage modulus increases gradually as the frequency increases, high-frequency waves travel faster than those of low frequency. Waves of ultra-high frequency travel at a speed $(G_g/\rho)^{\frac{1}{2}}$ based on the glass modulus, and waves of very low frequency travel at the much lower equilibrium wave speed $(G_e/\rho)^{\frac{1}{2}}$. Although the difference between these limiting velocities may be enormous, the range of frequencies over which the change occurs is even more enormous, and the velocity is nearly constant in any given decade of frequencies.

The frequency-dependence of the attenuation constant is mainly due to the explicit factor ω; the slowness and the loss tangent change only very gradually with frequency. A wave of frequency ω penetrates to a depth of the order of $1/\alpha \sim 1/\omega$ before being absorbed. High-frequency waves move off very fast but disappear immediately. The slower low-frequency waves may penetrate into the medium to a great depth.

Problem: Plot graphs of c and α versus ω for (a) the Maxwell model, (b) the Kelvin-Voigt model, and (c) the power-law model, with p small. Notice that the three plots do not resemble one another very much. The power-law model is the one that resembles data for real polymers.

The wave motion produced by an arbitrary wall motion $v_0(t)$ can be found by first Fourier-analyzing the wall motion, i.e., representing it as a linear combination of sinusoidal motions. The wave produced by each component is then found, as just discussed. The wave motions at station y due to all components are then put back together to find the resultant signal at station y.

The same result, in the same form, is found by using Laplace transforms and taking the inversion contour to be the pure imaginary axis $s = i\omega$. Thus, when the wall motion is a pulse, we obtain

$$V(y,t) = \mathrm{Re}\,\frac{1}{\pi}\int_0^\infty \exp[i\omega(t-c^{-1}y) - \alpha y]d\omega.$$

Here we have already noted that the integral along the negative half of the axis is the complex conjugate of that along the positive half, and combined the two; Re means the real part.

The integrand is oscillatory and exponentially damped. In most ranges of frequency, the contribution from half a cycle cancels out the contribution from the other half, so it's hard to see how the integral can amount to anything at all. However, there may be a frequency ω_0 at which the phase angle $\omega(t-c^{-1}y)$ is stationary. If the integrand is effectively non-oscillatory in some interval around ω_0, the integral over that interval may be appreciable. The stationary point ω_0 is determined as the solution of the equation

$$0 = \frac{d}{d\omega}[\omega(t-c^{-1}y)]$$
$$= t - c^{-1}y - c^{-1}y\,\frac{d(\log c^{-1})}{d(\log \omega)}\,.$$

Now,

$$\log c^{-1} = \tfrac{1}{2}\log|J^*| + \log(\cos\tfrac{1}{2}\delta) + \tfrac{1}{2}\log\rho.$$

If the loss angle is small then the term involving it is exceedingly small, so

$$\frac{d(\log c^{-1})}{d(\log \omega)} = \frac{1}{2}\,\frac{d(\log|J^*|)}{d(\log \omega)} = -\tfrac{1}{2}p \qquad \text{(say)}.$$

Here $-p$ is the slope on a log-log plot of $|J^*|$ at the relevant frequency. Although we do not know what frequency that may be, we assume that the slope is small everywhere. The equation for ω_o is

$$c(\omega_o) = (y/t)(1-\tfrac{1}{2}p).$$

Thus, for given values of y and t, ω_o is given to a first approximation as the frequency at which the phase velocity is y/t. Given a first approximation, the value of p there can be determined. A second approximation is then obtained as the frequency at which the phase velocity is $(y/t)(1-\frac{1}{2}p)$, with the estimated value of p. Because c changes so gradually, the second approximation may be considerably different from the first, but if p changes only gradually then little or no adjustment will need to be made after the second approximation.

To obtain the so-called second approximation directly, we note that with p small, the equation is

$$\log c = \log(y/t) - \tfrac{1}{2}p,$$

and since $\frac{1}{2}p$ is the slope of a plot of $\log c$ versus $\log \omega$, then

$$\log(y/t) = \log c + \left.\frac{d(\log c)}{d(\log \omega)}\right|_o [\log(\omega_o e) - \log \omega_o].$$

The term in brackets, equal to unity, has been included in order to make it more obvious that the right-hand member of the equation is the value of $\log c$ at the frequency $\omega_o e$. Thus, ω_o is lower by a factor of e than the frequency at which the phase velocity is y/t:

$$c(\omega_o e) = y/t.$$

We note that if y/t is larger than the maximum phase velocity $(G_g/\rho)^{\frac{1}{2}}$, there is no stationary point. Later we will show that in fact there is no signal at a point y further away than $c_g t$; the leading edge of the signal travels at a large finite speed.

Let the phase velocity at the stationary point be c_o. Then the exponent of the integrand is

$$-\tfrac{1}{2}pi\omega y c_o^{-1} - i\omega y(c^{-1}-c_o^{-1}) - \omega y c^{-1}\tan(\tfrac{1}{2}\delta).$$

Here we have used the equation defining ω_o to substitute for t in terms of c_o.

The final term, $-\omega y c^{-1}\tan(\frac{1}{2}\delta)$, accounts for the attenuation. If only frequencies in the vicinity of ω_o are important, then we can replace c and δ by their values at ω_o without committing much of an error. Furthermore, we recall that for a power-like compliance, the loss angle is approximately $\delta = \frac{1}{2}p\pi$, where p is the same log-log slope encountered previously. Assuming further that this is small, then the attenuation term finally becomes $-\omega y c_o^{-1} p\pi/4$.

The term $-i\omega y(c^{-1}-c_o^{-1})$ accounts for dispersion, i.e., the spreading of the signal due to the fact that waves of different frequencies have different velocities. If we neglect this term, then the phase is no longer stationary at any frequency; the integration is then easy to perform, but the result is silly.

(Problem: Check this.) Consequently, we will try a linearized approximation. In the standard method of stationary phase we would treat $c^{-1} - c_o^{-1}$ as a multiple of $\omega - \omega_o$, obtain a quadratic exponent, and do the integration. Here, however, we expect to get more accurate results if we linearize on a logarithmic scale. Consequently, we use the approximation

$$\begin{aligned} c^{-1} - c_o^{-1} &= c_o^{-1} \frac{d(\log c^{-1})}{d(\log \omega)}\bigg|_o (\log \omega - \log \omega_o) \\ &= -\tfrac{1}{2}pc_o^{-1} \log(\omega/\omega_o). \end{aligned}$$

At this point the exponent has become

$$-i\omega\tfrac{1}{2}pyc_0^{-1}[1 - \log(\omega/\omega_0)] - \omega\tfrac{1}{2}pyc_0^{-1}\tfrac{1}{2}\pi.$$

If we now make the change of variable $\omega' = \omega\tfrac{1}{2}pyc_0^{-1}$, then we obtain

$$V(y,t) \sim \operatorname{Re} \frac{1}{\pi} (\tfrac{1}{2}pyc_0^{-1})^{-1} \int_0^\infty \exp i\omega'[\log \omega' + i\tfrac{1}{2}\pi - \log(\omega_0'e)]d\omega' .$$

Now recalling the function $I(Z_0)$ defined earlier, we see that the present integral yields the same function:

$$V(y,t) \sim (\tfrac{1}{2}pyc_0^{-1})^{-1}I[\omega_0\tfrac{1}{2}pyc_0^{-1}].$$

The result begins to look more like that found previously for a pure power-law if we recall that $yc_0^{-1} = t/(1-\tfrac{1}{2}p) \cong t$:

$$V(y,t) \approx (\epsilon t)^{-1}I(\epsilon t\omega_0),$$

where $\epsilon = \tfrac{1}{2}p$. Letting $\Omega(c)$ be the inverse of $c(\omega)$, so that $\Omega(c)$ is the frequency at which the phase velocity is c, and recalling that $e\omega_0$ is defined as the frequency at which the phase velocity is y/t, we finally obtain

$$V(y,t) \sim (\epsilon t)^{-1}I[\epsilon te^{-1}\Omega(y/t)].$$

<u>Problem</u>: Verify that for a pure power-law compliance, with p small, this is the same as the result found earlier.

Problem: Verify the evaluation of the integral in terms of $I(Z_o)$.

e. The elastic precursor.

Although the main part of the pulse travels with a speed approximately equal to $(\rho J(t))^{-\frac{1}{2}}$, there is some small disturbance travelling a great deal faster than this. As the disturbance first enters a previously undisturbed region, the only part of the compliance that can possibly be relevant is its initial value J_g, so the initial disturbance runs ahead of the main pulse with a large velocity $(\rho J_g)^{-\frac{1}{2}}$.

To find out a little about this precursor wave, we consider the behavior of the transform $\exp(-ys/\sqrt{\rho s\bar{J}})$ for large s (small t). $\bar{J}(s)$ is given asymptotically for large s by the transform of the Taylor series expansion of $J(t)$,

$$s\bar{J}(s) \sim J(0) + J'(0)s^{-1} + \cdots .$$

With the notation J_g for the initial value and J'_g for the initial slope, then

$$(s\bar{J})^{\frac{1}{2}} \sim J_g^{\frac{1}{2}}[1 + (s\tau)^{-1} + \cdots],$$

where $\tau = 2J_g/J'_g$ is a very small time computed from the small initial value and large initial slope. Then the transform is given by $\exp(-ys/c_g)\exp(-y/c_g\tau)$, where $c_g = (\rho J_g)^{-\frac{1}{2}}$. The inverse of this is

$$\delta(t-yc_g^{-1})\exp(-y/c_g\tau) = \delta(t-yc_g^{-1})\exp(-t/\tau).$$

Thus, the initial response to a Dirac delta in wall velocity is a Dirac delta disturbance. As a measure of the strength of this disturbance at a station y, we can use its time integral, $\exp(-y/c_g\tau)$. This is to be compared with the time integral of the whole signal, unity (the value of the transform at $s = 0$). We see that although initially all of the momentum is in the precursor, within a negligible time there is hardly any left. For this reason the precursor is very hard to measure, and it requires very sophisticated techniques to detect it at all.

f. Long-time behavior.

After a time so long that $J(t)$ is near its equilibrium value J_e (for a solid) or is close to $(t+T)/\eta_o$ (for a fluid), we can predict what the pulse will be like without doing any mathematics. First, for a fluid, we can expect that the pulse will be asymptotically the same as for an ideal viscous fluid with viscosity η_o. Second, for a solid, we can expect that the pulse will be moving at the uniform speed $c_e = (J_e\rho)^{-\frac{1}{2}}$. It will also be spreading out, with a width probably proportional to $t^{\frac{1}{2}}$; the total viscosity (the integral of $G(t) - G_e$) probably has something to do with the width.

To find out whether or not any of this is right, we can use the expansion of $\bar{J}(s)$ in powers of s, for small s. Recall that the coefficients in the expansion involve the limiting value or slope of J, the total viscosity, the mean relaxation time, and, to a higher order than will be needed, higher moments. If we use this expansion in the transform $\bar{V}(y,s)$ then we should be able to perform the inversion by a saddle-point integration, assuming that y and t are large.

Problems.

1. Follow the procedure just outlined to show that the long-time behavior of a pulse in a fluid is indeed independent of everything but the steady-shearing viscosity.

2. Try to do the saddle-point integration for a solid as well. This is tricky because you will get zero for an answer unless you keep $(y-c_et)/t^{\frac{1}{2}}$ or some such thing fixed as y and t approach infinity. Find out what happens if you try to do the problem in whatever way seems most straightforward.

Additional Problems.

1. In defining the basic response functions, we have spoken of applying a unit step in shear or stress to a thin layer of material. Since signals do not travel with infinite velocity, it is not possible to apply a homogeneous deformation to all parts of a body simultaneously. Consider a layer of material of

thickness h, bonded to a rigid support along the plane $y = h$. At time zero the other side of the layer, $y = 0$, is subjected to a unit displacement tangential to itself. This causes a motion with velocity $v(y,t)$ in the layer. (a) Determine the transform of $v(y,t)$. (b) From the transform, determine the limiting value of the displacement at large times, $u(y,\infty)$. (c) Show that $v(y,t)$ can be expressed in terms of the pulse response function for a semi-infinite body, $V(y,t)$, as

$$v(y,t) = V(y,t) - V(2h-y,t) + V(2h+y,t) - V(4h-y,t) + \cdots .$$

(To do this, use a geometric series expansion $(1-e^{-x})^{-1} = \sum e^{-nx}$ at an appropriate place, and recall one of the shifting theorems.) Explain what is going on, in physical terms. (d) Estimate the amount of time that must elapse before the slab can be regarded as motionless and homogeneously deformed.

2. A slab or layer of thickness h is subjected to a sinusoidal shearing oscillation, so that the particle displacement $u(y,t)$ satisfies $u(0,t) = 0$ and $u(h,t) = A\exp(i\omega t)$. We have assumed that if the layer is sufficiently thin, then the deformation can be regarded as a sinusoidal simple shearing, $u(y,t) = (Ay/h)\exp(i\omega t)$. (a) Prove that this is approximately correct if $\omega h < c(\omega)$, where $c(\omega)$ is the phase velocity of a sinusoidal plane wave with frequency ω. That is, the deformation is approximately homogeneous if the time required for a shear wave to cross the slab is less than $1/\omega$. (b) Since the slab does have mass, the shearing stress on the surface $y = h$ is not exactly $G^*(A/h)\exp(i\omega t)$, the result obtained by neglecting the mass. One might expect that an additional stress would be required, equal to the rate of change of momentum of all of the material of the slab (per unit area). Estimate this additional stress by calculating the rate of change of momentum under the assumption that the displacement varies linearly. (c) Show that this correction is not quite right, by comparing it with the lowest-order correction obtained from the exact solution. Observe that the latter correction is, however, purely inertial as expected. (d) At sufficiently high frequencies, the slab is essentially motionless except in a thin layer near $y = h$. What is a "sufficiently high frequency"? What is the thickness of the boundary layer?

3. A fluid is at rest in a channel of width $2L$. It adheres to the boundaries $y = \pm L$. At time zero a body force per unit volume F in the x-direction is applied. As a consequence, the fluid begins to move in the x-direction with velocity $v(y,t)$. (a) Determine $\overline{v}(y,s)$. (b) From the transform, verify that after a long time the fluid moves just as an ideal viscous fluid with viscosity η_0 would. (c) Show that after a long time, the total displacement $u(y,t)$ is the same as for an ideal viscous fluid that started moving not at time zero, but at the time $t = -T$, where T is the mean relaxation time of the material. (d) If the force F is removed at time t_0 and the fluid is allowed to come to rest, the final displacement $u(y,\infty)$ is the same as for an ideal viscous fluid. Prove this. (e) Comparing a viscoelastic fluid and an ideal viscous fluid with the same steady-shearing viscosity, if the force F is applied to both at the same instant, the viscoelastic fluid surges ahead of the viscous one, but when the force is removed, it loses what it has gained. Explain why, in physical terms.

4. Set up the same problem, for flow in a tube of circular cross-section rather than a channel, and obtain the transform of $v(r,t)$.

5. With the same fluid in the same channel as in Problem 3, suppose that the force per unit volume along the channel direction is a sinusoidally oscillating force $F \exp(i\omega t)$. Consider the steady-state oscillation of the fluid. (a) Show that if the solution for the case of an ideal viscous fluid is known, the solution for a viscoelastic fluid can be obtained from it by substituting the complex viscosity $G^*(\omega)/i\omega$ for the viscosity of the ideal viscous fluid. (b) Find the velocity profile for low-frequency oscillations. What should the frequency be low in comparison to, in order to use a low-frequency approximation? (c) Show that at high frequencies the fluid moves back and forth as a rigid block (i.e., with no shearing) except in thin layers next to the walls. (d) Show that at high frequencies the largest r.m.s. velocity is not that of the rigid block in the middle, but occurs close to the wall. By how large a factor can the velocity near the wall exceed that in the middle?

6. Set up Problem 5 for flow in a tube rather than a channel. Express the transform $\overline{v}(r,s)$ in terms of Bessel functions.

CHAPTER V

STRESS ANALYSIS

We now consider some problems involving small, gradually changing deformations. The stresses and their changes over long periods are the matters of main interest. Deformations can also be of interest in themselves if the structure creeps, i.e., the deformation continues to change for a long time.

Problems of the same kind are considered in classical elasticity theory. Indeed, when a solid body under given static loads or displacements has reached equilibrium (i.e., no further changes of stresses or displacements are taking place), the viscoelastic solution is the same as the classical elastic solution, with elastic moduli equal to the equilibrium values of the viscoelastic stress-relaxation moduli. The similarity with elasticity theory goes much deeper, however. In many cases, viscoelastic stress analysis problems can be converted directly into elasticity problems by applying the Laplace transform.

1. Quasi-Static Approximation.

We use the quasi-static approximation, in which the inertial terms in the momentum equation are ignored. Thus, instead of seeking solutions of the momentum equation,

$$\rho\ddot{u}_i = \rho f_i + \sigma_{ij,j},$$

we consider the equilibrium equation instead, even though the displacements u_i will generally be changing in time:

$$\sigma_{ij,j} = -\rho f_i.$$

(Here ρ is the mass density, which can be regarded as equal to the prescribed initial density in small deformations. $\underline{f}$, with components f_i $(i = 1,2,3)$ is the body force per unit mass, ordinarily gravity. The stress component σ_{ij} is the

i^{th} component of the force per unit area on a surface element whose normal is in the jth coordinate direction. $\sigma_{ij,j}$ means $\partial\sigma_{ij}/\partial x_j$, where x_j is the jth coordinate. We use the summation convention: whenever a letter subscript appears exactly twice in a monomial, as j does in $\sigma_{ij,j}$, there is a missing summation sign, and a sum over all values of the subscript is taken for granted.)

In using the quasi-static approximation we do not claim that the rate of change of momentum is actually negligible. Whenever boundary forces or displacements are changed, a disturbance propagates into the body at a large but finite speed (loosely, the speed of sound). This wave motion is reflected at boundaries and may soon become random and incoherent. It would still be present even so, if there were no dissipative mechanism. However, because of stress relaxation (and heat conduction, which we have not considered), waves in viscoelastic materials are damped out very quickly, as we have seen in the particularly simple case of plane waves. The boundary disturbance is soon translated into a change in static deformation, plus heat. In using the quasi-static approximation we do not exactly neglect the momentum changes taking place during this process; we simply <u>ignore</u> the whole business, and pretend that at any instant the body is in equilibrium with the concurrent boundary data. This is justifiable if changes in boundary conditions or changes due to stress relaxation are taking place slowly enough that there is no appreciable change during the amount of time required for decay of a wave motion.

In classical elasticity theory, the quasi-static approximation converts a problem with time-dependent boundary conditions into a set of time-independent problems, numbered with a parameter t. In viscoelasticity theory, matters are not quite so simple. Even with constant (i.e., step) boundary conditions and omission of time derivatives in the momentum equation, problems remain truly time-dependent because the stress is determined by the history of the deformation and not solely by the present deformation. We will be dealing with integro-differential equations. This sounds very complicated, but we will find that, in fact, viscoelasticity problems are practically the same as elasticity problems.

2. Stress-Strain Relations.

In elasticity theory we assume that when displacements u_i and displacement gradients $u_{i,j}$ are sufficiently small, each component of stress is a linear function of the displacement gradients:

$$\sigma_{ij} = c_{ijkl}u_{k,l}.$$

If the material is isotropic, the matrix of stiffness coefficients c_{ijkl} reduces to an especially simple form, and the relation becomes

$$\sigma_{ij} = \delta_{ij}\lambda u_{k,k} + \mu(u_{i,j} + u_{j,i}),$$

where δ_{ij} is the Kronecker delta.

In viscoelasticity theory we assume that if a deformation is applied as a step at time zero, so that the displacement gradient history has the form $H(t)u_{i,j}$, then the stress response is of the form

$$\sigma_{ij}(t) = c_{ijkl}(t)u_{k,l}.$$

The stress-relaxation moduli $c_{ijkl}(t)$ are zero at negative times, and presumably positive and decreasing at positive times, being discontinuous at time zero.

Just as in the one-dimensional case, we partly assume and partly deduce that the stress response to a deformation varying in time is equal to a superposition of the stress responses to each incremental deformation:

$$\sigma_{ij}(t) = \int_{-\infty}^{t} c_{ijkl}(t-t')du_{k,l}(t') = c_{ijkl} * du_{k,l}.$$

As in elasticity theory, we find that if the material is isotropic then there are only two independent stress-relaxation moduli, and the relation is

$$\sigma_{ij} = \delta_{ij}\lambda * du_{k,k} + \mu * d(u_{i,j} + u_{j,i}).$$

The Lamé constants λ and μ are replaced by functions of time, and multiplication is replaced by convolution.

Problems:

1. Prove that if an infinitesimal rigid rotation of a body or of the observer's frame of reference produces no stress, then $c_{ijkl}(t) = c_{ijlk}(t)$. Conclude that

$$c_{ijkl} * du_{k,l} = c_{ijkl} * d\epsilon_{kl},$$

where $\underline{\epsilon}$ is the strain, $\epsilon_{ij} = \frac{1}{2}(u_{i,j} + u_{j,i})$.

2. Prove that if mass and linear momentum are conserved, if there are no couples except the couples of forces, and if there is no angular momentum except the moment of linear momentum, then, conservation of angular momentum implies that $\sigma_{ij} = \sigma_{ji}$. Prove that if arbitrary (small) displacement gradients can be produced without violating the conservation of angular momentum, then $c_{ijkl}(t) = c_{jikl}(t)$.

3. Explain the material symmetry argument which leads to the simplified form of stress-strain relation quoted for the isotropic case.

4. Show that if the material is isotropic, then $c_{ijkl}(t) = c_{klij}(t)$.

3. Simplest Deformations of Isotropic Materials.

Earlier we have considered only shearing deformations, and have described the response in terms of a single modulus $G(t)$. As an exercise, and to demonstrate the similarity with elasticity theory, let's see why this agrees with what has just been said about the stress-strain relation for isotropic materials.

Consider a time-dependent simple shearing deformation,

$$u_1 = x_2\kappa(t), \quad u_2 = u_3 = 0.$$

All displacement gradients except $u_{1,2} = \kappa(t)$ are zero. Hence, the isotropic stress-deformation relation yields $\sigma_{ij} = 0$ for all components of stress except

σ_{12} ($\equiv \sigma_{21}$), which is

$$\sigma_{12} = \mu * d\kappa .$$

Thus, $\mu(t)$ is just another notation for the shear modulus $G(t)$.

The second modulus $\lambda(t)$ accounts for the major part of the response to volume changes. Let $v(t)$ be the dilation, i.e., the volume change per unit initial volume, $u_{i,i}$. In a pure radial compression or expansion, the displacement field has the form

$$u_i = x_i v(t)/3 .$$

Then $u_{i,j} = \delta_{ij} v(t)/3$. The stress-strain relation yields

$$\sigma_{ij} = \delta_{ij} K * dv,$$

where K is the <u>bulk modulus</u>, defined by

$$K(t) = \lambda(t) + \frac{2}{3} \mu(t).$$

Operationally, it would be better to say that $\lambda(t)$ is defined in terms of the bulk modulus and the shear modulus by this relation.

Let s_{ij} and e_{ij} be the deviatoric components of stress and strain, defined by

$$s_{ij} = \sigma_{ij} - (1/3)\sigma_{kk}\delta_{ij}, \qquad e_{ij} = \epsilon_{ij} - (1/3)\epsilon_{kk}\delta_{ij} .$$

Then $s_{ii} = e_{ii} = 0$. The general stress-strain relation for isotropic materials can be decomposed into shearing response and volume response as

$$s_{ij} = 2G * de_{ij} \quad \text{and} \quad \sigma_{kk} = 3K * d\epsilon_{kk} .$$

The inverses of these relations are

$$e_{ij} = \tfrac{1}{2} J * ds_{ij} \qquad \text{and} \qquad \epsilon_{kk} = (1/3)B * d\sigma_{kk}.$$

Here $J(t)$ is the familiar shear compliance, and $B(t)$ is the bulk compliance.

Problem: Write the integral equation that relates $B(t)$ to $K(t)$.

Problem: Torsion.

Consider the torsion of a rod of arbitrary uniform cross-section. First consider an elastic material. By using the linearity of the problem and using dimensional analysis, show that the relation between the twisting moment M and the angle of twist per unit length, θ, is of the form $M = GS\theta$, where S depends on the shape and dimensions of the cross-section and might conceivably depend on the ratio K/G (it does not). Express M in terms of the history of θ, and θ in terms of the history of M, for a viscoelastic material.

4. Simple Tension: Basic Approximation Methods.

In simple tension, all stress components are zero except some tensile stress component, $\sigma_{11} = \sigma(t)$, say. Then it follows from the stress-strain relations that $\epsilon_{ij} = 0$ if $i \neq j$, that $\epsilon_{22} = \epsilon_{33}$, and that

$$\epsilon_{11} - \epsilon_{22} = \tfrac{1}{2} J * d\sigma \qquad \text{and} \qquad \epsilon_{11} + 2\epsilon_{22} = (1/3)B * d\sigma.$$

Let us write $\epsilon_{11} = \epsilon(t)$ for the extension in the direction of tension. By eliminating ϵ_{22} from the relations above, we obtain

$$\epsilon = (J/3 + B/9) * d\sigma = D * d\sigma \quad \text{(say)}.$$

The function $D(t)$ is the tensile creep compliance. We write the inverse of this relation as

$$\sigma = E * d\epsilon.$$

The function $E(t)$, which replaces Young's modulus of elasticity or the Trouton viscosity of a fluid, is the <u>tensile relaxation modulus</u>.

If a rod is subjected to a strain history $\epsilon(t)$ by means of tensile forces on its ends, with no traction on its sides, then there is a lateral contraction $-\epsilon_{22}(t)$. For an elastic material the ratio $-\epsilon_{22}/\epsilon$ is constant in time. For a viscoelastic material, Poisson's ratio is not constant. Indeed, we find that

$$-\epsilon_{22} = \tfrac{1}{2}\,\epsilon - (1/6)B * d(E*d\epsilon) = \nu * d\epsilon \qquad \text{(say)},$$

where $\nu(t)$ is defined by

$$\nu = \tfrac{1}{2}\,H - (1/6)B * dE.$$

Let us consider some approximate methods of determining the stress response $\sigma(t)$ accompanying a given extension history $\epsilon(t)$. Since we use the same methods in much more complicated problems, it is worth while to examine these methods first in a context that is not algebraically or geometrically complicated. Here the problem boils down to finding E, given G and K.

We do not try to guess what the response to an input $\epsilon(t)$ might be, separately for each new input. Instead, we seek to determine the approximate stress response $\sigma(t) \cong E_a(t)$ only in the simplest case, that of a unit step input. More complicated inputs are then treated by superposition: $\sigma = E_a * d\epsilon$. If we get the step response half-way right, then the response to more complicated inputs is likely to be given quite accurately because in the superposition, errors tend to smooth out and cancel.

In the relatively simple case of a step input $\epsilon(t) = \epsilon H(t)$, most of the deformation occurs at or near time zero. Of course, the lateral contraction $-\epsilon_{22}$ will generally vary with time, but we can expect that its behavior will be more or less step-like, with only gradual changes except at short times (short in comparison to the time of observation). If we pretend that <u>all</u> of the deformation occurs at time zero, we obtain

$$\sigma(t) = 2G(t)(\epsilon-\epsilon_{22}) \qquad \text{and} \qquad \sigma(t) = 3K(t)(\epsilon+2\epsilon_{22}).$$

These relations are contradictory unless

$$-\epsilon_{22}/\epsilon = \tfrac{1}{2}\, H(t) - \tfrac{1}{2}[K(t)/G(t) + (1/3)]^{-1}.$$

Evidently the best choice of the "constant" ϵ_{22} depends on the time at which the stress is to be evaluated. If the right-hand side is indeed fairly constant over recent times, the method is self-consistent, and we obtain $\sigma(t) \cong E_a(t)$, where the approximate modulus is

$$E_a(t) = \left[\frac{1}{3G(t)} + \frac{1}{9K(t)}\right]^{-1}.$$

Note that if G and K were constants, this would be the exact expression for Young's modulus in terms of G and K. We call this the <u>quasi-elastic</u> approximation.

<u>Problem</u>: Derive the same result by assuming that $G(t)J(t)$, $B(t)K(t)$, and $D(t)E(t)$ are all approximately unity at all times.

The quasi-elastic approximation is widely useful in quasi-static problems. To determine the response to a <u>step</u> input, we simply find the corresponding response for an elastic material, and then substitute time-dependent relaxation moduli or compliances for the elastic constants. For example, if we designate $-\epsilon_{22}$ as the "response", we first determine the elastic response $-\epsilon_{22} = \nu\epsilon$, i.e., we determine the expression for Poisson's ratio ν in terms of the elastic constants G and K. Then for a viscoelastic material, we claim that the approximate response $-\epsilon_{22}(t) = \nu_a(t)\epsilon$ for a <u>step</u> input $\epsilon(t) = \epsilon H(t)$ is obtained by replacing G and K in the elastic expression by $G(t)$ and $K(t)$:

$$\nu_a(t) = \tfrac{1}{2}\, H(t) - \tfrac{1}{2}[K(t)/G(t) + (1/3)]^{-1}.$$

Finally, we take the response to an arbitrary input to be what is obtained by superposition, $-\epsilon_{22} = \nu_a * d\epsilon$.

We next consider the synchronous modulus approximation. We suppose that $K(t)$ is merely some constant multiple of $G(t)$: $K(t) = cG(t)$. But if that is the case, then Poisson's ratio $\nu(t)$ is exactly constant rather than merely approximately so, and for a step input, $\epsilon(t) = \epsilon H(t)$, the lateral strain ϵ_{22} is exactly a step. Thus the assumptions used in the quasi-elastic approximation are satisfied exactly, so the results are exact. The synchronous modulus approximation does not give a different result, but only justifies the assumptions used in obtaining the quasi-elastic approximation.

This is an approximation because it is not really true that the bulk modulus is a constant multiple of the shear modulus. What is true, is that the bulk modulus is usually so much larger than the shear modulus that there aren't going to be any volume changes anyway, if the material can find some other way to deform, so it doesn't much matter what we say that the bulk modulus is. Typically (for an amorphous polymer), the initial values K_g and G_g are of the same order or magnitude, but G relaxes to an equilibrium value $G_e = 10^{-4} G_g$, say, while K does not relax nearly so much, say $K_e = 10^{-1} K_g$. Thus K is a great deal larger than G except at very short times.

Problem: Does a rod held at constant extension get thinner, or fatter, as time progresses?

One approximation suggested by this behavior is to treat K as infinite, i.e., treat the material as incompressible, unless, of course, the boundary conditions require or suggest that important volume changes will take place. In the present case of simple tension, there is no compelling reason why the volume should change very much. If we take the bulk compliance B to be zero, i.e., negligible, then $D = J/3$, so $E = 3G$. This is exactly the same as what would have been obtained by setting $1/K = 0$ in the quasi-elastic approximation. We can also view incompressibility as a special case of synchronous moduli, with $K(t) = \infty G(t)$. The quasi-elastic approximation is exact if the material is incompressible, and is still useful in other cases. Nevertheless, incompressibility is still an important method of approximation in itself, because it simplifies matters so much.

Let us use the notation $\bar{\bar{F}} = s\bar{F}$ for the s-multiplied Laplace transform of a function F. Then the transforms of the relations

$$\sigma_{ij} = c_{ijkl} * du_{k,l} \qquad \text{and} \qquad \sigma_{ij,j} = -\rho f_i,$$

with zero initial conditions, are

$$\bar{\bar{\sigma}}_{ij} = \bar{\bar{c}}_{ijkl}\bar{\bar{u}}_{k,l} \qquad \text{and} \qquad \bar{\bar{\sigma}}_{ij,j} = -\rho\bar{\bar{f}}_i .$$

Assuming that the boundary conditions can also be transformed in the same way, we see that the transforms satisfy what is formally an <u>elasticity</u> problem. This is the exact version of the <u>correspondence principle</u> relating viscoelasticity problems to elasticity problems (see E. H. Lee, Quart. Appl. Math. <u>13</u>, 183(1955)).

In the present problem, the correspondence principle means that $\bar{\bar{E}}(s)$ and $\bar{\bar{\nu}}(s)$ are given exactly in terms of $\bar{\bar{G}}(s)$ and $\bar{\bar{K}}(s)$ by substituting these operational moduli for the corresponding elastic constants in the expressions given by elasticity theory:

$$\bar{\bar{E}}(s) = \left[\frac{1}{3\bar{\bar{G}}(s)} + \frac{1}{9\bar{\bar{K}}(s)}\right]^{-1}, \quad \bar{\bar{\nu}}(s) = \tfrac{1}{2} - \tfrac{1}{2}\,[\bar{\bar{K}}(s)/\bar{\bar{G}}(s) + (1/3)]^{-1}.$$

If analytical expressions for G and K are known, it may be possible to obtain E and ν exactly from these relations, but usually this is not the case. However, we can use approximate transform and inversion formulas here. For example, assuming that all of these functions are power-like, we can use Schapery's Rule, $\bar{\bar{F}}(s) \cong F(1/2s)$. If we use this formula to transform G and K and then use it again in the inversion to find E and ν, the result is exactly that given by the quasi-elastic approximation.

Of course, we can also use the more exact transform formulas that take into account the log-log slope of the function. In the present case this gives rather complicated expressions for E and ν unless G and K have the same log-log slope. In that case, however, G and K are proportional (synchronous moduli), so the quasi-elastic solution is exact and we don't need to worry about transform inversion.

Because $K(t)$ relaxes much less than $G(t)$ does, it is fairly realistic to approximate the bulk modulus by a constant. The approximation of <u>elastic bulk response</u> does not simplify matters much in the present problem. However, to test the quasi-elastic approximation in a case involving moduli that are definitely not synchronous, let us consider the case K = constant, $G(t) = Ct^{-p}/(-p)!$, which is much like real behavior in some important respects. With $\bar{\bar{K}} = K$ and $\bar{\bar{G}} = Cs^p$, then

$$\bar{\bar{E}}(s) = 9K[1 + (3K/Cs^p)]^{-1}.$$

For large s (small t) this gives

$$\bar{\bar{E}}(s)/9K = 1 - (3K/Cs^p) + (3K/Cs^p)^2 - + \cdots .$$

Hence

$$E(t)/9K = 1 - t^p(3K/Cp!) + t^{2p}(3K/C)^2/(2p)! - \cdots .$$

On the other hand, the quasi-elastic approximation gives

$$E_a(t)/9K = [1 + 3K/G(t)]^{-1} = 1 - t^p(3K/C)(-p)! + t^{2p}(3K/C)^2(-p)!^2 - \cdots .$$

To first order in p, $(-p)!$ is the same as $1/p!$, $(-p)!^2 = 1/(2p)!$, etc. Hence, the quasi-elastic approximation gives practically the exact solution if p is small, at least for small t. For large t (small s), $\bar{\bar{E}}$ is asymptotic to $3Cs^p$, so $E(t) \sim 3G(t)$, as if the material were incompressible. This occurs because K becomes infinitely large in comparison to $G(t)$ as time progresses; of course, K is not changing, but the effect of G becoming very small is the same as that of K being very large. It is for this reason that we can say that the material is "incompressible" when K is much larger than G. The quasi-elastic approximation gives the same result, of course.

Problem: Simple Extension.

Suppose that a thin disc or wafer of material is bonded along its faces to two rigid plates. Let the x_1-direction be perpendicular to the faces. Then if the plates are squeezed closer together or pulled further apart, the displacement will be mainly in the x_1-direction, except close to the edge of the disc: $u_1 = x_1 \epsilon(t)$, $u_2 = u_3 = 0$. The normal force per unit area required to produce this deformation is $\sigma = M * d\epsilon$. Express the bulk longitudinal modulus $M(t)$ in terms of the shear and bulk moduli. Let the compliance form of the relation be $\epsilon = N * d\sigma$. Find an approximate expression for $N(t)$ in terms of $G(t)$ and $K(t)$, assuming that K is much larger than G. Can the assumption of incompressibility be used if $K >> G$?

5. The Correspondence Principle.

Let us consider boundary-value problems in some generality. We need to specify boundary conditions, as functions of time, that would make the problem well-set as a quasi-static elasticity problem. Here we add an additional restriction because of the technique of solution that is to be used: The kind of data that is specified at a given boundary point $\underline{x}$ is the same at all times. For example, if the displacement component u_i at $\underline{x}$ is specified for all time, we can evaluate the transform $\bar{u}_i$ immediately from the data, and thus regard $\bar{u}_i(\underline{x},s)$ as given. On the other hand, if u_i is specified from time zero to time t_1, and then the traction T_i is specified from time t_1 onward, we do not have enough data to evaluate either $\bar{u}_i$ or $\bar{T}_i$ immediately. Transform techniques can still be used, but Wiener-Hopf methods will be involved, and the great simplicity of the correspondence principle will be lost.

We have already remarked that applying the Laplace transform to the equilibrium equation and the stress-deformation relation yields equations that are formally the same as in elasticity theory. With the preceding stipulation about the boundary conditions, the whole problem is the same as an elasticity problem. The thing to do, if possible, is to find the solution of the elasticity problem in a book, substitute s-multiplied transforms for all variables and elastic constants, and then perform the inversion.

If the elastic solution is known, all of the work is in the transform inversion. It will not usually be possible to carry out the inversion exactly. Consequently, it's worth taking a little care ahead of time to simplify the transform inversion.

First, if a problem has complicated boundary conditions, do not try to solve it all in one piece. The equations are all linear, so we can superpose solutions of simple problems to get the solution of a complicated problem. For example, it is standard practice to set the body force term ρf_i equal to zero when solving problems. If the body force is in fact important, then the solution obtained by omitting it is part of the full solution nevertheless. To complete the full solution, it would be necessary to solve a second problem in which all data is zero except the body force. The complete solution is the sum of the two parts. (Caution: Each part must be a well-set problem by itself.)

Second, try to set up each partial problem so that wherever the prescribed data is not zero, it all varies in proportion to a single function of time. For example, suppose that displacements or tractions are prescribed as zero over part of the boundary, and on the remainder, the traction is prescribed as $\sigma_{ij}(\underline{x},t)n_j(\underline{x}) = T_i(\underline{x})F(t)$, so that although the traction may vary with position, all tractions vary in the same proportion in time. In the case of non-zero displacement boundary conditions, we would like to deal with proportional surface displacements, $u_i(\underline{x},t) = U_i(\underline{x})u(t)$. If given data $F(\underline{x},t)$ varies with $\underline{x}$ and t together in a non-separable way, then separate it nevertheless: On a patch near $\underline{x}_o$, replace $F(\underline{x},t)$ by $F(\underline{x},t_o)F(\underline{x}_o,t)/F(\underline{x}_o,t_o)$, and off this patch, replace F by zero. Of course, we will then be doing only part of the problem, and will have to treat other patches separately. This is not a very clean way of doing the space and time separation, but is only offered as an illustration that such separation is possible.

Third, now assuming that only a single function of time $f(t)$ is involved in the boundary data, solve the problem only for a step input $f(t) = H(t)$.

Fourth, ask yourself what it is that you want to know about the solution. There is no point in inverting the transforms of all sorts of miscellaneous functions that you aren't even interested in. Suppose, for example, that the displacement $u(t)$ at some particular point is the main thing you want to know.

Now, with all prescribed data varying in proportion to a single function $f(t)$, we can regard $f(t)$ as the input. The function $u(t)$ is the output. These are related linearly. Let $U(t)$ be the output corresponding to a step input $f(t) = H(t)$. Then the output corresponding to a general input is obtained by superposition: $u = U * df$. If this is only part of the original problem, we will have to add similar contributions to u from the other inputs.

If a problem has been simplified to this point, then what we will need to find from elasticity theory is the transform $\bar{\bar{U}}(s)$ of the response to a unit step input. Let the elastic solution for the corresponding variable be $U_e(c)$, say, where c stands for whatever elastic constants may be involved. By replacing these constants by the s-multiplied transforms $\bar{\bar{c}}(s)$ of the corresponding viscoelastic response functions, we get $\bar{\bar{U}}(s) = U_e(\bar{\bar{c}}(s))$. We now use Schapery's formula to obtain

$$U(t) \cong \bar{\bar{U}}(1/2t) = U_e(\bar{\bar{c}}(1/2t)).$$

But, with another application of the same approximation, $\bar{\bar{c}}(1/2t)$ is equal to the viscoelastic response function c evaluated at t. Thus, we obtain

$$U(t) \cong U_e(c(t)).$$

In other and simpler words, we simply replace elastic constants in the elastic solution by viscoelastic response functions, to get the viscoelastic response to a step input. This is just the quasi-elastic approximation, discussed previously in the special case of simple tension. Of course, one might wish to use a more accurate method of inversion, but the quasi-elastic method is attractive for its simplicity and its usually good accuracy.

One word of caution: From a physical point of view, we expect the quasi-elastic method to be accurate if the response is step-like. From a mathematical point of view, we know how to generalize this notion a little; we expect the approximate transform inversion to be accurate if the function involved is power-like. When the response $U(t)$ has been obtained by the quasi-elastic method, inspect it. If it has any prominent interior maxima or minima, or otherwise fails to resemble a power, then it is not likely to be entirely accurate, so a more refined method should be tried, to be on the safe side.

6. Example: Flat-Headed Punch.

In contact problems, such as the indentation of a half-space by a rigid spherical punch, the region of contact changes in time, so a given boundary point may be stress-free up until some unknown time and then later have a prescribed displacement, after coming into contact with the indenter. Since the boundary condition is not of the same type at all times, such problems cannot be treated entirely directly by transform methods. Moreover, such problems are not linear, even in elasticity theory. The equations are linear, but the boundary conditions are not.

The situation is much simpler in the case of a flat-headed punch. In that case the region of contact is known from the outset, and the problem is linear. Let F be the total indenting force, and let D be the depth of indentation, which is constant throughout the known contact region. To find the relation between F and D, we first consider the elastic case. Evidently D will be proportional to F, and inversely proportional to the stiffness G (or perhaps K, but we will account for this). To make matters definite, suppose that the contact area is circular, with radius R. Then, putting in factors of R where needed to make the dimensions come out right, we find that the solution in the elastic case must be of the form

$$D = (F/RG)f(\nu),$$

where $f(\nu)$ is some pure number depending on Poisson's ratio.

For a viscoelastic material, the transformed solution is $\bar{\bar{D}} = R^{-1}f(\bar{\bar{\nu}})\overline{\overline{JF}}$. Then 1) if the material is incompressible, $\nu = \frac{1}{2}$ and $D = R^{-1}f(\frac{1}{2})J * dF$. 2) If K and G have the same time-dependence, then ν is constant, and $D = R^{-1}f(\nu)J * dF$. 3) In the quasi-elastic approximation, $D = R^{-1}(fJ) * dF$, where fJ, at time t, is $f(\nu(t))J(t)$.

If the punch is removed at time t_0, say, then the recovery is <u>not</u> given exactly by setting $F = 0$ for $t > t_0$ in the preceding results. For, once the punch is no longer in contact, the <u>kind</u> of boundary condition satisfied in the "contact" region is changed from a prescribed-displacement condition to a zero-traction condition, and the preceding results, based on the implicit assumption that there is contact, are no longer valid. Nevertheless, we can get an estimate of the mean depth of the dent during recovery by continuing to use the relation already derived.

7. <u>Example: Tube Under Internal Pressure.</u>

Let us consider a thick-walled tube with internal and external radii a and b, respectively. We suppose that it undergoes a purely radial deformation, with displacement $u(r,t)$, under the action of an excess pressure $P(t)$ in the cavity. The outer surface is fixed: $u(b,t) = 0$.

The displacement equation of equilibrium under no body force takes the form

$$G * d\nabla^2\underline{u} + (K + G/3) * d\underline{\nabla}(\underline{\nabla}\cdot\underline{u}) = \underline{0}.$$

In the present problem, with $\underline{u}$ of the form $\underline{i}_r u(r,t)$, this is a second-order linear ordinary differential equation for u, so far as the r-dependence is concerned. The time-dependence doesn't matter at first because we start off by considering the corresponding elastic problem, in which the convolution operation reduces to multiplication by a constant. Now, it saves a little work to remember that pure two-dimensional expansion $u = r$ must be a solution because it is a homogeneous deformation. Also, remember the solution $u = 1/r$, which is a fundamental singular solution in two dimensions. (Notice that both of these fields satisfy $\nabla^2 \underline{u} = \underline{0}$ and $\underline{\nabla}(\underline{\nabla} \cdot \underline{u}) = \underline{0}$ separately, so it doesn't matter what the elastic constants are.) Since the general solution is a linear combination of two linearly independent solutions, we have

$$u = C_1 r + C_2/r.$$

With the boundary condition $u(b,t) = 0$, we obtain

$$u(r,t) = u_a(t)\,\frac{(b/r) - (r/b)}{(b/a) - (a/b)},$$

where $u_a(t) = u(a,t)$.

As sometimes happens, this solution does not involve the elastic constants. Consequently, the same relation is valid for the viscoelastic case. Making a longer story of it to illustrate the general method, we first interpret u and u_a in the elastic solution as s-multiplied transforms of the corresponding functions in the viscoelastic case, and then perform the transform inversion. Here this gets us back where we started.

We can now calculate all stresses exactly, in terms of the history $u_a(t)$, by using the viscoelastic stress-strain relation.

Problem: Do so. Notice that for the kind of deformation considered here, $\epsilon_{rr} = \partial u/\partial r$, $\epsilon_{\theta\theta} = u/r$, and all other strain components are zero.

Since the internal pressure $P(t)$ is $-\sigma_{rr}(a,t)$, we find that the relation between $P(t)$ and $u_a(t)$ is

$$a(R^2-1)P(t) = 2[K + (G/3)(1+3R^2)] * du_a,$$

where $R = b/a$.

<u>Problem</u>: Express $u_a(t)$ in terms of the history of $P(t)$ by using the quasi-elastic approximation on this relation.

To understand this relation a little better, let's look at some special cases. Suppose that K is a great deal larger than G. First consider the case of a thin-walled tube, in which R is approximately unity. Then the relation between P and u_a becomes $(b-a)P(t) = K * du_a$. Since the volume of the tube wall, per unit of length in the axial direction, is $V = \pi(b^2-a^2)$, and the change of volume is $dV = -2\pi a du_a$, this is simply $P = -K * dV/V$. Thus, the bulk response controls the deformation. Recall that the bulk <u>longitudinal</u> modulus is $M = K + 4G/3$. In terms of it, we can write the relation for a thin-walled cylinder more exactly as $P = M * du_a/(b-a)$.

At the opposite extreme $R \to \infty$, we are in effect dealing with a cylindrical cavity in an infinite medium. We obtain $aP = 2G * du_a$. In this case the total volume change is spread out very thinly, so the bulk response is unimportant.

A point of main interest in the solution is the hoop stress $\sigma_{\theta\theta}$, particularly at the inner boundary $r = a$. There could be some worry that the tube might crack when pressurized, and it seems reasonable to suppose that if there is a crack, it will be radial, starting from the inside. For an elastic material, the hoop stress at the inner boundary is

$$\sigma_{\theta\theta}(a,t) = -\frac{K + G/3 - R^2G}{K + G/3 + R^2G} P(t).$$

From this we can immediately say what will happen in the viscoelastic case. We replace K and G by the corresponding stress-relaxation moduli. Then, we see that if the tube wall is not very thick, and $K \gg G$, the hoop stress is practically equal to $-P(t)$. In particular, the stress is compressive, so no cracking should be expected. In keeping with what was found earlier, the response is mainly a pressure-volume relation. Next consider the opposite extreme $R = \infty$. In this case the tensile hoop stress is equal to $P(t)$. Thus, a very thick-walled tube may crack if suddenly pressurized. Since the conclusions do not actually involve the moduli in these two extreme cases, they are directly valid for an arbitrary pressure history.

However, the assumption that $K \gg G$ is not very realistic at short times. Suppose that $K_g = 2G_g$ and $R = 2$, i.e., $b = 2a$. Then for a unit step in pressure, the hoop stress is instantaneously equal to $+5/19$, and it is tensile. Assuming that G relaxes much more rapidly than K does, then the hoop stress is ultimately close to -1. Thus, because of viscoelastic effects, the hoop stress is a tensile stress immediately after loading, although it later becomes compressive.

Problems.

1. Writing the relation between inflation pressure and hoop stress as $\sigma_{\theta\theta} = L * dP$, what is $L(t)$, according to the quasi-elastic approximation? Let t_o be the time at which L vanishes, $L(t_o) = 0$, if there is such a time. Show that in pressurization at a constant rate starting at time zero, the maximum (tensile) hoop stress is $\dot{P} \int_0^{t_o} L(t)dt$. Show that if the pressurization is removed after the hoop stress has reached equilibrium, the hoop stress becomes even more compressive.

2. Consider a pressurized tube as in the preceding problem, but with the condition $\sigma_{rr}(b,t) = 0$ at the outer boundary. Show that the radial and hoop stresses are proportional to the concurrent value of the inflation pressure $P(t)$ at each instant, whether the material is elastic or viscoelastic. Show that in contrast, the axial stress depends on the history of pressurization, in general; $\sigma_{zz} = 2(R^2-1)^{-1}\nu * dP$. Discuss the radial displacement caused by a step in pressure. Include consideration of the special cases $R = 1$ and $R = \infty$.

8. Incompressible Materials.

In the problem concerning pressurization of a tube with its external boundary fixed, we could have assumed that if K is always much larger than G, then the material is effectively incompressible. We would have obtained the very simple solution that there is no deformation, and the stress is an isotropic pressure $P(t)$ at every point. Although there is some truth to this, of course it overlooks everything of importance. In contrast, in the problem of pressurization with the external boundary free, nothing of any importance is missed by treating the material as incompressible. As we shall now show, it always simplifies matters a great deal if the material can be treated as incompressible, but one should bear in mind that the validity of this idealization depends on what problem is being considered.

If we suppose that the bulk compliance is zero, then $u_{i,i} = 0$. The mean normal stress is indeterminate if the dilation is too small to notice and the bulk modulus is too large to comprehend, so the stress takes the form

$$\sigma_{ij} = -p\delta_{ij} + 2G * d\epsilon_{ij},$$

where p is not specified directly in terms of the deformation.

Let us first consider pure displacement boundary-value problems. The displacement equation of equilibrium takes the form

$$\underline{\nabla} p = G * d\nabla^2 \underline{u},$$

since $\underline{\nabla} \cdot \underline{u} = 0$. Given a displacement field $\underline{u}$, there exists a pressure p satisfying this equation only if $G * d\underline{\nabla} x \nabla^2 \underline{u} = \underline{0}$. Thus $\underline{\nabla} x \nabla^2 \underline{u} = J * d\underline{0} = \underline{0}$. Hence, the displacement satisfies

$$\underline{\nabla} \cdot \underline{u} = \underline{0}, \qquad \nabla^2 \underline{\nabla} x \underline{u} = \underline{0},$$

and $\underline{u} = \underline{U}$ (given) on the boundary.

Notice that all material properties and all history dependence have dropped out. For incompressible materials, in pure displacement boundary-value problems, the displacement field is determined at each instant by the boundary conditions at that instant, and the displacement field is the same as in elasticity theory. The problem is also the same as a Stokes flow problem for an incompressible, ideally viscous fluid, if we interpret $\underline{u}$ as velocity rather than displacement.

The usefulness of this result would at first sight appear to be drastically restricted because the problem is not well-set unless the boundary displacements satisfy the condition of zero volume change, $\Delta V = \oiint \underline{U} \cdot \underline{n} \, dS = 0$. However, this is easy to fix. If ΔV is not zero, then let the displacement field be $\underline{u}' = (\Delta V/3V)\underline{x} + \underline{u}$, a pure dilation of the right amount to account for the total volume change, plus a remainder $\underline{u}$ that now satisfies the equations previously given. The new field $\underline{u}$ is equal to $\underline{U} - (\Delta V/3V)\underline{x}$ on the boundary.

Now let us consider pure traction boundary-value problems. In this case we would like to express all equations to be solved in terms of the stresses, p and the deviatoric stress s_{ij}. We have

$$p_{,i} = s_{ij,j}, \quad 2\epsilon_{ij} = u_{i,j} + u_{j,i}, \quad u_{i,i} = 0, \quad s_{ij} = 2G * d\epsilon_{ij}.$$

We need to eliminate u_i and ϵ_{ij}. Now, it is well-known that there exists a displacement field u_i satisfying the strain-displacement relation only if the strain satisfies the compatibility conditions, $e_{rip}e_{sjq}\epsilon_{ij,pq} = 0$. Hence, the deviatoric stress must satisfy the same conditions. If it does, the strain field $\epsilon_{ij} = \frac{1}{2}J * ds_{ij}$ derived from it will furnish a displacement field. Moreover, $u_{i,i}$ will be zero if $s_{ii} = 0$, and only then. Hence, the problem expressed entirely in terms of stress is

$$p_{,i} = s_{ij,j}, \quad e_{rip}e_{sjq}s_{ij,pq} = 0, \quad s_{ii} = 0,$$

with traction specified on the boundary:

$$(-p\delta_{ij} + s_{ij})n_j = T_i \qquad \text{(say)}.$$

Again, in this kind of problem, all material properties and all history-dependence have dropped out. For incompressible materials, in pure traction boundary-value problems, the stress field is determined at each instant by the boundary conditions at that instant, and the stress field is the same as in elasticity theory.

Problem: With synchronous moduli (constant Poisson ratio), in displacement boundary-value problems the displacement field is determined at each instant by the boundary displacements at that instant, and in traction boundary-value problems the stress field at each instant is determined by the surface tractions at that instant alone. Solutions generally depend on the ratio of moduli. Prove these statements.

CHAPTER VI

THERMAL EFFECTS

Thermal effects are important in viscoelasticity in the same way that they are important in the special cases of classical elasticity and fluid dynamics. But in addition, in viscoelasticity theory there are really drastic temperature effects of a kind never considered in the classical theories. The mean relaxation time of a material depends very strongly on the temperature. Generally, the higher the temperature is, the quicker the material relaxes or complies. This kind of temperature-dependence is, of course, absent from the theories that involve no relaxation time.

The equilibrium (elastic) moduli depend on temperature, but this is as uninteresting in viscoelasticity as it is anywhere else. Effects of thermal expansion are important for the same reasons as in classical theories. We will accordingly pay little attention to equilibrium properties and thermal expansion, and instead concentrate on the effects of temperature-dependence of the mean relaxation time.

1. The Time-Temperature Shift Factor.

Viscoelastic effects in high polymers are most prominent when these materials are in the rubbery state, rather than the glassy state. The equilibrium shear modulus of an ideal rubber is directly proportional to absolute temperature. Thus, unlike metals, elastomers become stiffer when heated. However, this refers to equilibrium behavior. Since stress relaxes much faster when the material is heated, a viscoelastic material may very well seem less stiff at higher temperature, depending on what one is doing to it.

The rate of stress relaxation is determined by the material's own internal clock. So far as the material knows, stress relaxation always proceeds at the same rate. However, the hotter the material is, the faster everything moves around on a molecular scale. In one tick on the material's clock, the amount of observer's time that has elapsed is $a(\theta)$, say, an amount that decreases rapidly as the temperature θ increases.

Let $G(t;\theta)$ be the stress relaxation modulus at a constant temperature θ, and let $G(t) = G(t;\theta_o)$ be the modulus at a reference temperature θ_o. Then, taking $a(\theta_o) = 1$,

$$G(t;\theta) = (\theta/\theta_o)G[t/a(\theta)].$$

The modulus has the same shape, and G/G_e runs through the same values, at all temperatures. The main effect of temperature is on the time scale $a(\theta)$. Recalling the definitions of the steady-shearing viscosity and the mean relaxation time (of a fluid), we see that both are directly proportional to $a(\theta)$, aside from the relatively unimportant factor θ/θ_o:

$$\eta_o(\theta) = \int_0^\infty G(t;\theta)dt = a(\theta)(\theta/\theta_o)\int_0^\infty G(t')dt' = a(\theta)(\theta/\theta_o)\eta_o(\theta_o),$$

$$\eta_o(\theta)T(\theta) = \int_0^\infty tG(t;\theta)dt = \eta_o(\theta_o)T(\theta_o)(\theta/\theta_o)a^2(\theta).$$

Indeed, by dividing the second expression by the first, we find that the ratio of mean relaxation times, $T(\theta)/T(\theta_o)$, is exactly $a(\theta)$. If we should choose a reference temperature at which the mean relaxation time were unity, then $a(\theta)$ would be equal to $T(\theta)$ at all temperatures.

The scaling factor $a(\theta)$ depends strongly on the temperature, especially near the glass transition temperature θ_g. This is the temperature at which, in principle, the thermodynamic equilibrium behavior of the material changes from glassy to rubbery. Below the glass transition, the material is a glass, a hard amorphous solid. When the material is above the glass transition, it is more like ordinary rubber. For example, room temperature is above the glass transition temperature of anything you would call "rubber", but it is below the glass transition temperature of plexiglass (polymethyl methacrylate). The glass transition is a second-order change of phase, in contrast to first-order changes such as the transition from ice to water.

Near the glass transition, it seems that the temperature-dependence of

$a(\theta)$ is given pretty well for all amorphous polymers by the Williams-Landel-Ferry formula,

$$\log_{10} a(\theta) = -16.14 \frac{\theta - \theta_0}{56 + \theta - \theta_0} + C.$$

Here C and θ_0 depend on the particular material. θ_0 is about equal to the glass transition temperature. At high temperatures, where $a(\theta)$ levels off according to the WLF formula, it appears that $a(\theta)$ may actually continue to decrease as $a(\theta) = A \exp(-B\theta)$, at the rate of about a factor of 10^4 for each 100° change in temperature.

We see that when $\theta - \theta_0$ increases from -10° to $+10^\circ$, the right-hand side of the WLF formula decreases by about 6, so $a(\theta)$ decreases by a factor of a million when the temperature is increased by twenty degrees. The amount of stress relaxation that takes place in one minute at the higher temperature would require twenty years at the lower temperature.

As mentioned earlier, the equilibrium modulus increases with temperature. However, suppose we don't wait for equilibrium, but measure the stress one minute after application of a strain step. At temperatures below the glass transition, there is negligible relaxation in one minute, so the one-minute modulus is the same as the initial value G_g. That is why G_g is called the glass modulus. Now, as the temperature increases, the amount of stress relaxation that can take place in one minute also increases, so the one-minute modulus decreases very rapidly with a rise in temperature. Thus, from a real and natural point of view, the material is getting softer as the temperature increases. If the material is finally brought to a high enough temperature that it can relax essentially to equilibrium in one minute, then at that temperature and higher, the one-minute modulus is the equilibrium modulus, which increases slowly with temperature.

The glass transition temperature is sometimes crudely determined as a temperature in whose neighborhood the viscosity suddenly changes by a factor of a million or so. Since the viscosity is proportional to the relaxation time $a(\theta)$, we see from the WLF formula that the viscosity will increase by a factor of a million in a change from 10° above the glass transition to 10° below it. However, this is not an actual discontinuity, so it does not give a precise determination of θ_g. Nothing else gives a precise determination of θ_g, either. It would certainly be possible to recognize the temperature at which a second-order change in equilibrium properties occurs (for example a discontinuity in the modulus) if it were possible to obtain equilibrium data. However, as θ is lowered toward θ_g, the relaxation time grows so enormous that no one has time to wait for equilibrium.

The "softening" effect of a decrease in the relaxation time is very clear in connection with the dynamic modulus. Since the transform of $G(t/a)$ is $a\bar{G}(as)$, and the s-multiplied transform is $\bar{\bar{G}}(as)$, then the dynamic modulus is $G^*(a\omega)$, (times θ/θ_o, but let that go). Since G^* decreases as the frequency decreases, we see that at a fixed frequency ω, G^* decreases rapidly as θ increases. For example, in a range of time (or reciprocal frequency) for which $G(t) = Ct^{-p}/(-p)!$ and thus $G^*(\omega) = C(i\omega)^p$, then at a constant temperature different from the reference temperature we would have

$$G^*(\omega,\theta) = (\theta/\theta_o)C\omega^p[a(\theta)]^p\exp(ip\pi/2).$$

Even if p is small, $a(\theta)$ goes down so rapidly with an increase of temperature that the modulus will decrease noticeably.

With any one experimental method it is difficult to measure G (or G^*) over a range of times (or frequencies) differing by a factor of more than a thousand, say. If observations can be made at times from one second to one hour, then to find out what's going on at times less than a second at room temperature, one simply cools the material enough that what was one second at room temperature becomes one hour at the lower temperature. To observe the very slow relaxation at times longer than an hour, speed it up by heating the material. Whenever you see data covering ten or twelve decades in time, from a microsecond to a week or two, it is likely that data from experiments at various temperatures have been combined into one curve by using the shift factor. Graphically, if G or $\log G$ is plotted against the logarithm of time, then since $\log t/a(\theta) = \log t - \log a(\theta)$, the time scaling becomes simply a bodily translation or shifting along the log time axis. The curve shifts to the left as temperature increases. If curves for various temperatures, but all in the same range of log t, are available, the various curves are shifted right or left until they all fit together into one curve, and the amount of shift that was necessary for each segment determines the value of $a(\theta)$ for the temperature at which that segment was observed.

Problem: Explain the temperature-dependence of the compliance $J(t;\theta)$ and the complex compliance $J^*(\omega;\theta)$. Sketch curves showing what J will look like over a limited period of time, for temperatures ranging from well below the glass transition to far above it.

Reading: Ferry, especially Chapter 11; articles by Leaderman and Tobolsky in Eirich's Rheology, Volume 2.

2. Example: Runaway.

Suppose that a slab of material is sheared sinusoidally. If the rate of work, $\sigma\dot{\kappa}$, is integrated over a complete cycle, we find that the work per cycle (per unit volume) is $\pi G_2\kappa_o^2$ or $\pi J_2\sigma_o^2$, where σ_o and κ_o are the stress amplitude and the shear amplitude, and G_2 and J_2 are the loss modulus and loss compliance. Multiplied by the number of cycles per second, $\omega/2\pi$, this gives the average rate of dissipation:

$$D = \frac{1}{2}\omega G_2\kappa_o^2 = \frac{1}{2}\omega J_2\sigma_o^2.$$

(Problem: Verify these statements.)

If the slab is insulated so that no heat can escape, this energy input will cause a rise in temperature. So far as the storage modulus is concerned, this temperature rise will definitely "soften" the material, unless the frequency is so tiny that the material is effectively in thermodynamic equilibrium at all times. Let us suppose that the loss modulus also decreases as the temperature increases.

If the shearing is carried out at a constant shear amplitude κ_o, then the stress amplitude gradually decreases as the material becomes hotter. Consequently, the rate of dissipation also decreases. The temperature rise will become slower and slower, and, if heat is allowed to flow out of the slab, the temperature will approach some steady-state value.

On the other hand, suppose that the stress amplitude σ_o is held constant. Then as the material heats, the compliance increases, the shear amplitude increases, and so the rate of dissipation also increases. There is a feed-back effect. The hotter the material is, the faster the temperature goes up. The situation is unstable, and there will be some sort of a catastrophe. There may be trouble even if heat is allowed to flow out, because it may not be able to flow out fast enough.

To get a more definite idea of what will happen, we need to consider the energy equation,

$$\rho\dot{e} = -q_{i,i} + \sigma_{ij}\mathring{u}_{i,j}.$$

We suppose that the internal energy density e is determined by the history of deformation and temperature. In a quasi-steady-state cyclic oscillation, the histories of deformation and temperature are characterized by their cycle-averaged values and the amplitudes and frequency of their fluctuations. If we consider only the average rate of increase of internal energy over a complete cycle, then the term $\rho\mathring{e}$ can be replaced by $C\theta_t$, where θ_t is the rate of increase of average temperature and C is a specific heat. C might conceivably depend on σ_o, probably depends on ω, and very likely depends on θ, but we will treat it as a constant.

For the heat flux we use Fourier's law, $q_i = -K\theta_{,i}$. We treat the conductivity K as constant, although it undoubtedly depends on the average temperature at least. In the present problem, if we let the direction normal to the slab be the y-direction, then the average temperature will depend only on y and t, so $-q_{i,i}$ will take the form $K\theta_{yy}$.

For the average rate of work $\sigma_{ij}\mathring{u}_{ij}$, we use the rate of dissipation D

discussed previously. We consider the case in which the stress amplitude is constant, so the expression for D in terms of J_2 is convenient. We finally obtain

$$C\theta_t = K\theta_{yy} + \tfrac{1}{2}\omega\sigma_o^2(\theta_o/\theta)J_2[a(\theta)\omega].$$

The methods to be used can be carried out just as well if J_2 and $a(\theta)$ are empirically determined functions, given graphically or numerically. However, to be explicit, let us suppose that J_2 has a power-law form, $J_2(\omega) = j\omega^{-p}$, and that $a(\theta) = A\exp(-B\theta)$. To simplify matters, let us omit the factor θ_o/θ.

Let the boundary of the slab be $y = \pm h$. We scale y with respect to h, so that the boundary is $y = \pm 1$ in the new variable. We also scale t and θ in such a way as to get rid of as many parameters as possible, and translate the temperature origin so that the new dimensionless temperature ϕ is initially zero. The equation can thus be brought into the form

$$\phi_t = \phi_{yy} + \tfrac{1}{2}g^2\exp(\phi).$$

Here g is a dimensionless parameter.

(<u>Problem</u>: Carry out the scaling, and find out what g is.)

Let us first consider the case of insulated boundaries. In that case the temperature remains uniform at all times if it is uniform initially, so the equation yields

$$t = (2/g^2)\int_0^{\phi} \exp(-\phi)\,d\phi.$$

Of course, the integration can be carried out explicitly, but visualize a case in which the integrand is given in the form of data. The main thing to notice about the integral is that it converges as $\phi \to \infty$, so the temperature diverges to infinity within a finite time t_∞ (equal to $2/g^2$ in the present case). If we were talking about a nuclear reactor, this would be the time when it blows up.

If we let heat escape through the boundary, then perhaps there will be no instability. With $\phi = 0$ at $y = \pm 1$, let's look for a steady-state distribution $\phi(y)$. By multiplying the equation by ϕ_y and integrating, we obtain

$$\phi_y^2 + g^2 \exp(\phi) = g^2 \exp(\phi_o).$$

With a symmetrical distribution, so that $\phi_y(0) = 0$, the parameter ϕ_o is the temperature at $y = 0$. This is evidently the maximum temperature. From the equation, evidently ϕ_y approaches $\pm g \exp(\phi_o/2)$ as y goes to $\mp\infty$. Thus, the function defined by this equation increases linearly as y comes up from $-\infty$; its slope then decreases to zero at its maximum at $y = 0$, and it then goes down symmetrically on the other side, finally decreasing linearly for $y \to \infty$. Thus in an infinite medium there could be a self-sustained temperature peak which would not diffuse away.

(Problem: Carry out the integration to find yg as a function of ϕ and ϕ_o.)

Since the maximum temperature is ϕ_o and the limiting slope is $g \exp(\phi_o/2)$, then if ϕ_o is large, ϕ will vanish at about $y = \pm (\phi_o/g)\exp(-\frac{1}{2}\phi_o)$. To make ϕ vanish at $y = \pm 1$, the maximum temperature must satisfy $\phi_o \exp(-\frac{1}{2}\phi_o) = g$. But if g exceeds $2/e$, there is no solution.

(Problem: Use the exact solution to show that there is a critical value of g, above which there is no steady-state solution satisfying $\phi(\pm 1) = 0$.)

When g is greater than the critical value, there is certainly an instability, and we can expect that the temperature will grow indefinitely large. The result in the homothermal case suggests that the temperature in the middle of the slab will diverge to infinity in a finite time. For g less than the critical value, there is a steady-state solution. Presumably heat can diffuse away through the boundaries fast enough, for g small, that there is no runaway.

In a problem with realistic data, it looks like one should concentrate on finding out 1) the critical value g_{cr}, 2) the steady-state maximum temperature

ϕ_o, for $g < g_{cr}$, and 3) the burn-out time t_∞, for $g > g_{cr}$. The first two items are easy to obtain with the present simple model. The third is not so easy. However, to obtain at least an estimate, we can do this: Pretend that the temperature is $\phi = \phi_o(t)(1-y^2)$. Put this into the equation; it will certainly not satisfy the equation pointwise. Integrate over y. This yields

$$(4/3)\phi_o'(t) = -4\phi_o(t) + g^2 \exp \phi_o(t)\cdot I[\phi_o(t)].$$

Here $I(\phi_o)$ is the integral, expressible in terms of the error function,

$$I(\phi_o) = \phi_o^{-1/2} \int_0^{\phi_o^{1/2}} \exp(-y^2)\,dy.$$

It is equal to unity at $\phi_o = 0$ and goes down like $\frac{1}{2}(\pi/\phi_o)^{1/2}$ for ϕ_o large.

The model equation can now be integrated to give t as a function of ϕ_o:

$$t = (4/3) \int_0^{\phi_o} [g^2 I(\phi) \exp \phi - 4\phi]^{-1} d\phi.$$

(Problem: Show that if g is less than some critical value, the integrand has a simple pole at some value ϕ_o^*, so the temperature will tend to this value as $t \to \infty$. Show that if g exceeds the critical value, the integral from zero to infinity converges, so the temperature goes to infinity in a finite time t_∞. Sketch the behavior of ϕ_o as a function of t for various values of g. Pay particular attention to the behavior for g very slightly larger than the critical value.)

(Reading: R. A. Schapery and D. E. Cantey, AIAA Journal 4, 255 (1966). N. C. Huang and E. H. Lee, J. Appl. Mech. March 1967, 127.)

3. Variable-Temperature Histories.

As we have seen, the main effect of temperature variation is to speed processes up or slow them down. Let ξ be the time measured on the material's internal clock. The amount of observer's time dt that elapses during a change $d\xi$ in material time is $a(\theta)d\xi$, or, $d\xi = \phi(\theta)dt$, where $\phi(\theta) = 1/a(\theta)$. We have alluded to some experimental results that show the usefulness of this notion in cases in which the temperature is held constant during an experiment. We now consider the effect of varying the temperature during the experiment.

Following Morland and Lee (Trans. Soc. Rheo. 4, 233 (1960)), we suppose that if the temperature varies, viscoelastic processes still go forward according to a material time defined by

$$\xi(t) = \int_0^t \phi[\theta(t')]dt'.$$

To begin with a specific case, consider a single-step shearing history, $\kappa(t) = \kappa_0 H(t)$. We suppose that the stress response is

$$\sigma(t) = [\theta(t)/\theta_0]G(\xi)\kappa_0.$$

The explicit factor θ/θ_0 should more generally be replaced by $G_e(\theta)/G_e(\theta_0)$. It is intended to account for the fact that the equilibrium modulus is temperature-dependent. As we shall see later, this factor should probably be replaced by some sort of temperature-relaxation effect.

Let us suppose that during this single-step shear history, the temperature is initially the reference temperature at which $\phi(\theta_0) = 1$. Suppose that at time t_1 it is suddenly lowered to a value θ_1 well below the glass transition temperature, where $\phi(\theta_1)$ is extremely small. At time t_2 the temperature is raised to θ_0 again.

When the material is at the temperature θ_0, material time and observer's time progress at the same rate. At the temperature θ_1, however, the material's clock is stopped, very nearly. The material time varies as

$$\xi(t) = \begin{cases} t & t \leq t_1, \\ t_1 & t_1 \leq t \leq t_2, \\ t_1+(t-t_2) & t_2 \leq t. \end{cases}$$

Then the stress is

$$\sigma(t)/\kappa_o = \begin{cases} G(t) & t \leq t_1, \\ [G_e(\theta_1)/G_e]G(t_1) & t_1 \leq t \leq t_2, \\ G(t_1+t-t_2) & t_2 \leq t. \end{cases}$$

In the interval when the material is "frozen", the stress remains constant at a value close to what it had just before freezing. (There is probably a delayed effect coming from the temperature change, which we neglect.) At the thawing time, the stress begins to relax again. So far as the material is concerned, the interval from t_1 to t_2 just never happened.

(Problem: Suppose that in the preceding example, θ_1 is a very high temperature, at which the mean relaxation time is much shorter than t_2-t_1. What is the stress for $t > t_2$?)

In a variable shearing history, a shear step $d\kappa(t')$ at time t' makes a contribution to the stress at time t equal to

$$\delta\sigma(t) = (\theta/\theta_o)G[\xi(t)-\xi(t')]d\kappa(t'),$$

and the total stress at time t is a superposition of these parts. Let us suppose that the temperature history is again the step history considered previously. For times up to t_1, the stress is given by the usual expression in terms of the strain history. However, for times in the glassy interval $t_1 \leq t \leq t_2$, the stress increments due to strain changes before t_1 all remain constant, and stress increments due to strain changes after t_1 are also non-relaxing. Thus, the stress is

$$\sigma(t) = \sigma(t_1+) + G_g(\theta_1)[\kappa(t)-\kappa(t_1+)].$$

The behavior of the material is elastic, with a residual stress.

At time t_2, relaxation begins again. For times greater than t_2, the stress is

$$\sigma(t) = \int_0^{t_1} G(t-t_2+t_1-t')d\kappa(t') + \int_{t_2}^{t} G(t-t')d\kappa(t')$$

$$+ G(t-t_2)[\kappa(t_2)-\kappa(t_1)].$$

The first term represents the continuing relaxation of what was a "residual stress" during the glassy interval. Its rate of relaxation depends on how it was produced in the first place; the strain history up to time t_1 has by no means been "forgotten" while the material was glassy. The second term is the ordinary viscoelastic contribution from the deformation after thawing. The third term is the relaxation, beginning at the thawing time, of the total change of stress during the glassy interval. Everything about the strain history during the glassy interval has been forgotten except the total change of strain.

For a general history of strain and temperature, the stress is given by

$$\sigma(t) = [\theta(t)/\theta_0] \int_{-\infty}^{t} G[\xi(t)-\xi(t')]d\kappa(t').$$

With ξ as the variable, this is still a convolution, and its inverse is

$$\kappa(t) = \int_{-\infty}^{t} J[\xi(t)-\xi(t')]d[\theta_0\sigma(t')/\theta(t')].$$

Notice how the temperature has become enmeshed with the stress. Now, recall that the explicit θ/θ_0 term was put in as a fudge factor. Existing empirical data is isothermal, and it is just as consistent with this data to suppose that the correct form of the creep integral is

$$\kappa(t) = [\theta_0/\theta(t)] \int_{-\infty}^{t} J[\xi(t)-\xi(t')]d\sigma(t').$$

I do not know whether or not there is any data that would indicate which is correct, and I doubt that either is. Evidently there must be delayed effects of a temperature change in either relaxation or creep, so, abiding by the principle of maximum ignorance, I assume that there are effects both ways. At our present state of knowledge, the factor θ/θ_0 is either omitted when doing problems or it is put in wherever the analysis makes it most convenient. Fortunately, the changes in the equilibrium modulus are not the most important effect of temperature; of course, this is why no one has looked at it very much.

(Problem: Recall that the arithmetic mean is greater than the geometric mean. Regarding a Riemann sum as an arithmetic mean, show that

$$(1/t) \int_0^t \exp(B\theta)dt \geq \exp [(B/t) \int_0^t \theta dt].$$

Conclude that material time is always at least as large as the approximation to it obtained by using the time-average temperature.)

4. Example: Simple Tension of a Cooling Rod.

To show how complicated things become when the temperature varies both in time and in space, let us consider a problem that sounds simple but isn't. Let us suppose that an extending force F is applied to a rod of radius R at time zero Suppose that at time zero, the temperature of the rod is everywhere θ_1, but its surface $r = R$ is lowered to the reference temperature θ_0 at time zero and held there. The ends of the rod are insulated. The temperature is then a function of r and t only, at least away from the ends (we will not go into detail about how the load is applied). It satisfies the initial and boundary conditions

$$\theta(r,0) = \theta_1, \quad \theta(R,t) = \theta_0.$$

We neglect mechanical energy changes and dissipation, and thus suppose that the energy equation can be written as

$$C\theta_t = (K/r)(r\theta_r)_r.$$

The temperature can now be computed explicitly. This, at least, is a great simplification in comparison to the problem we would face if the work term in the energy equation were important.

We suppose that far enough away from the ends of the rod, the extension is uniform, a function $\epsilon(t)$ of time only. We omit consideration of thermal expansion; suppose that the material is incompressible. The tensile stress is then

$$\sigma(r,t) = \int_{0-}^{t} E[\xi(r,t) - \xi(r,t')]d\epsilon(t').$$

Here, the material time at the radius r is

$$\xi(r,t) = \int_{0-}^{t} \phi[\theta(r,t')]dt'.$$

The condition of equilibrium under the specified load F is then

$$\int_{0}^{R} \sigma(r,t)2\pi r dr = FH(t).$$

By combining the stress-strain relation with the equilibrium equation, we obtain

$$\int_{0-}^{t} E^{\#}(t,t')d\epsilon(t') = \sigma_0 H(t),$$

where $\sigma_0 = F/\pi R^2$ and where the effective modulus is defined by

$$E^{\#}(t,t') = (2/R^2) \int_{0}^{R} E[\xi(r,t)-\xi(r,t')]rdr.$$

Although the equation that $\epsilon(t)$ satisfies is not a convolution, it is still a Volterra equation.

To solve the problem exactly, with given functions $\phi(\theta)$ and $E(\xi)$, about all that is required is plenty of computer time; there is no conceptual difficulty. Before doing that, however, it might be a good idea to decide exactly what it is you want to know about the solution. In this connection, it is helpful to obtain at least a crude approximate solution, so you can see what you might need to be more precise about, if anything.

Just after time zero, the temperature is θ_1 everywhere except at the surface, so the rod extends uniformly according to the tensile compliance $D(\phi_1 t)$. The stress is σ_o almost everywhere. Right at the surface $r = R$, though, the modulus is $E(t)$ rather than $E(\phi_1 t)$, because the surface temperature is θ_o. With the quasi-elastic approximations $\epsilon(t) = D(\phi_1 t)\sigma_o \cong \sigma_o/E(\phi_1 t)$ and $\sigma(R,t) \cong E(t)\epsilon(t)$, we find that the stress at the surface is

$$\sigma(R,t) \cong \sigma_o E(t)/E(\phi_1 t).$$

Assuming that ϕ_1 is large, $E(\phi_1 t)$ relaxes much faster than $E(t)$ does, so the surface stress quickly becomes a good deal larger than σ_o. If $E(\phi_1 t)$ and $E(t)$ can both be approximated by the same power law $Ct^{-p}/(-p)!$ (after a little time has elapsed, so that both are somewhat below the glass modulus), then we obtain $\sigma(R,t) \cong \sigma_o \phi_1^p$. If $\theta_1 - \theta_o = 100^o$, then ϕ_1 might be about 10^4, but p is likely to be small, say $p = 0.1$, so ϕ_1^p is only about $2\frac{1}{2}$.

The surface stress seems to go up to about 2.5 times the nominal stress (with these figures) very quickly. It will then come back down, more slowly. As the lower temperature θ_o penetrates into the material, the annulus of stiff material on the outside grows thicker, and it will rapidly start supporting most of the load, both because it is stiffer and because a layer of given thickness has much bigger area near the outside than a layer of the same thickness has near the middle (the factor r in the integral defining $E^{\#}$). For a rough idea, let us suppose that there is a locus $r(t)$ such that the temperature is roughly θ_o for $r > r(t)$ and θ_1 for $r < r(t)$. As a further approximation on top of this, but using it, suppose that

$$E^{\#}(t,0) = \{1 - [r(t)/R]^2\}E(t) + [r(t)/R]^2 E(\phi_1 t).$$

Then, with $\sigma(R,t) = \sigma_o E(t)/E^{\#}(t,0)$, we can get an idea of how the surface stress will begin to decrease as the load-carrying (cool) region grows larger.

Problem: Let $r(t)$ be a locus of constant temperature, θ_m. What would be a good choice of θ_m? Determine $r(t)$ or its inverse, $t(r)$, either from the exact solution $\theta(r,t)$ or from the following approximate method (for small t): Take $\theta = \theta_1$ for $r < R-h(t)$ and, for $r > R-h(t)$, take $\theta = \theta_1 - (\theta_1-\theta_o)(r-R+h)^2/h^2$. Determine $h(t)$ by using either the exact equation for θ or the approximation $C\theta_t = K\theta_{rr}$, for h small. Notice that the equation defines a characteristic time $t_c = CR^2/K$. What is the physical significance of this time? Suppose that the material relaxes so slowly, even at the temperature θ_1, that $E(\phi_1 t_c) \cong E_g$. How would the rod behave in that case?

5. Thermal Expansion.

In elasticity theory, we choose some configuration of a body and call it "undeformed". Measuring strains from there, we express the stress as a function of the strain and the temperature:

$$\sigma_{ij} = -p(\theta)\delta_{ij} + \lambda(\theta)\epsilon_{kk}\delta_{ij} + 2\mu(\theta)\epsilon_{ij};$$

(for isotropic materials). Here we have made a Taylor series expansion with respect to the strain, and cut it off after terms of the first degree. If we also expand about a base temperature θ_o, and neglect second-degree terms again, we obtain

$$\sigma_{ij} = -p_o\delta_{ij} - p_o'(\theta-\theta_o)\delta_{ij} + \lambda_o\epsilon_{kk}\delta_{ij} + 2\mu_o\epsilon_{ij}.$$

The temperature variation makes no contribution to the deviatoric stress, since the material is isotropic: $s_{ij} = 2\mu_o \epsilon_{ij}$. It does affect the mean normal stress, however:

$$\sigma_{kk}/3 = -p_o - p_o'(\theta-\theta_o) + K_o \epsilon_{kk}.$$

Ordinarily, p_o stands for atmospheric pressure. The "stress" one talks about is guage stress, taking atmospheric pressure as zero on the guage. If we use the same symbol σ_{ij} for guage stress, then the term involving p_o disappears, but the term involving its derivative remains (unless we should choose a new reference configuration every time the temperature changes, which is possible only in homothermal cases). The coefficient p_o' is written as $K_o 3\alpha$, and α is called the linear thermal expansion coefficient. (3α is the <u>volume</u> thermal expansion coefficient.) We obtain

$$\sigma_{kk}/3 = K_o[\epsilon_{kk} - 3\alpha(\theta-\theta_o)].$$

By generalizing the preceding relation to viscoelastic rather than elastic response, we obtain

$$\sigma_{kk}/3 = K{*}d(\epsilon_{kk} - 3\alpha{*}d\theta),$$

(dropping the zero subscripts on quantities evaluated at θ_o). In terms of the bulk compliance B (compressibility), the inverse is

$$\epsilon_{kk} = B{*}d\sigma_{kk}/3 + 3\alpha{*}d\theta.$$

The function $\alpha(t)$ is the linear thermal expansion produced by a step change in temperature, $\theta(t)-\theta_o = H(t)$. Nothing much is known about it. Of course, $\alpha(\infty) = \alpha_e$ is the usual equilibrium value. Ferry quotes some experimental work by Kovacs (Ferry, Chapters 11 and 18) that has a bearing on it. Out of ignorance, we might decide to take α constant at its equilibrium value if there is a problem to solve and no data available. Notice, however, that with α constant (i.e. $\alpha(t) = \alpha_e H(t)$) then the thermal stress coefficient $K*d\alpha$ is $\alpha_e K(t)$. Thus if the volume response to a step change in temperature were instantaneous, the stress response would not be. Conversely, if we suppose that the thermal stress coefficient is constant at the value $\alpha_e K_e$, then $\alpha(t)$ is proportional to $B(t)$. Presumably neither of these simplifying assumptions is generally valid.

What people do, in order to sweep all of this under the rug, is to suppose that the bulk response is elastic, so that K, B, and α can all be treated as constants. In the many problems in which bulk response is not especially important, this is perfectly adequate and may even be a more precise model than one needs. However, one should realize that in the absence of data there is no way to be absolutely certain.

The stress-strain-temperature relation given above is strictly linearized with respect to temperature fluctuations. We can make an extrapolation that encompasses the most important non-linear effect of temperature variation by using Morland and Lee's reduced time (material time) $\xi(t)$ in place of ordinary time. The main known non-linear temperature effect still missing is the temperature-dependence of the equilibrium moduli, and with it, the transient changes in the moduli following a temperature step. Hopefully, these effects are not very important.

(Reading: Some interesting problems involving the production of residual stresses by the interaction of thermal expansion with non-uniform stiffening during cooling are discussed by Lee, Rogers, and Woo, J. Am. Cer. Soc., 48, 480 (1965), and Lee and Rogers, J. Ap. Mech., 32, 874 (1965). Muki and Sternberg (J. Appl. Mech. 28, 193(1961) were the first to make any progress in solving problems involving non-uniform material time.)

CHAPTER VII

LARGE DEFORMATIONS WITH SMALL STRAINS

The solution of a problem of finite deformation may require knowledge of material properties not embodied in the linear stress-relaxation moduli. Then again, it may not. In some problems of finite deformation, no material element is distorted very much even though displacements and rotations are large. In such cases we need to know only those material properties that entered into the description of response in infinitesimal deformations. However, if there are large geometry changes, it is necessary to take these into account in the kinematics.

We first show how to modify the linear viscoelastic equations for cases involving large translations and rotations (but small distortions). This is a fairly dry business involving the introduction of deformation gradients and finite strain measures. However, this is not to be avoided because bringing in the notion of the strain history for a finite deformation is the real ulterior motive for the present chapter. We then consider a few of the simplest possible examples involving such a formulation, partly for exercise, partly to build a bridge toward the discussion of viscoelastic flow problems, and partly to show that some fairly surprising results can be obtained.

1. Example: Simple Rotation.

Consider an initially quiet body, rotated through 90° at time zero, so that the position $\underline{x}$ of a particle initially at $\underline{X}$ becomes $x_1 = -X_2$, $x_2 = X_1$, $x_3 = X_3$. I'm not sure what is the most reasonable way to do the problem incorrectly, but suppose we claim that the displacements are $u_1 = x_1 - X_1 = -X_2 - X_1$ and so on; nothing wrong with that. In evaluating displacement gradients we run into a little problem. Perhaps $u_{1,1}$ should mean

$$u_{1,1} = \partial(x_1 - X_1)/\partial x_1 = 1,$$

or perhaps

$$u_{1,1} = \partial(x_1 - X_1)/\partial X_1 = -1.$$

If we evaluate displacement gradients the first way, we get a dilation $u_{i,i} = 2$, and if the second way, $u_{i,i} = -2$. From the constitutive equation $\sigma_{kk} = 3K{*}du_{i,i}$, we find that the mean normal stress is $2K(t)$, or perhaps $-2K(t)$. Of course, this is all nonsense. Turning a body through 90° certainly causes no volume change and no stress. Linear viscoelasticity theory is certain applicable, since we have not even distorted the body at all, but it is evidently a bad idea to try to use displacement gradients as measures of distortion if displacements are large.

2. Example: Torsion.

Consider the torsion of a long rod of radius R_o. Suppose that a particle initially at R, θ_o, Z moves so that at time t its coordinates are r, θ, z, where

$$r = R, \quad \theta = \theta_o + \psi(t)Z, \quad z = Z.$$

Here $\psi(t)$ is the angle of twist per unit length. The amount of shear of a material element in the azimuthal direction is $\kappa(r,t) = r\psi(t)$. If $R_o\psi(t)$ is small at all times, then no material element is ever distorted very much, so we are justified in using linear viscoelasticity theory in evaluating stresses, and indeed we have done so earlier, without comment. However, notice that if the rod is very long, $\psi(t)Z$ can be arbitrarily large even while $\psi(t)R_o$ is arbitrarily small. The angular displacement $\psi(t)Z$ of a particle need not be small; we can use linear viscoelasticity theory even though there are particles whose final positions are nowhere near their initial positions. However, we should not and need not claim that displacements and rotations are small, so we do not use the kinematics of infinitesimal deformations based on the use of displacement gradients.

Large deformations with small distortions are ordinarily handled, just as in the present case, by automatically doing the obvious things without making a federal case of it. However, it is sometimes convenient to have a theoretical apparatus that can be applied routinely.

3. Small Distortions with Large Rotations.

If a body is subjected to a deformation with infinitesimal displacement field $\underline{u}(\underline{X},t)$ and is, in addition, rotated by a finite amount, then the final position $\underline{x}(\underline{X},t)$ of a particle initially at $\underline{X}$ is

$$\underline{x}(\underline{X},t) = \underline{R}(t)[\underline{X} + \underline{u}(\underline{X},t)].$$

Here $\underline{R}$ is an orthogonal matrix, i.e. $\underline{R}^t\underline{R} = \underline{I}$, det $\underline{R} = 1$. In terms of components with respect to a cartesian coordinate system, the relation is

$$x_i = R_{iA}(X_A + u_A).$$

The stress components for the infinitesimal deformation are

$$\sigma^*_{AB} = c_{ABCD}{}^*du_{C,D}.$$

With the rotation superimposed, the stress components are

$$\sigma_{ij} = R_{iA}R_{jB}\sigma^*_{AB}.$$

Usually, even if a specified deformation can be decomposed into a small distortion and large rotation, we will not know this decomposition at the outset. Moreover, the best choice of the finite rotation is likely to be different at different particles. Consequently, it is convenient to re-write the relations above in a form that does not mention $\underline{u}$ or $\underline{R}$ explicitly.

Let $\underline{F}$ be the deformation gradient defined by $d\underline{x} = \underline{F}d\underline{X}$. In cartesian coordinates, $dx_i = F_{iA}dX_A$, so the components F_{iA} are the partial derivatives $x_{i,A}(\partial x_i/\partial X_A)$. Given the final position $\underline{x}(\underline{X},t)$ in terms of the particle label $\underline{X}$, these components can be calculated immediately. If the mapping of a fiber element $\underline{dX}$ onto $\underline{dx}$ is mainly a rotation, then $\underline{F}$ is nearly an orthogonal transformation $\underline{R}$. For example, with a deformation of the form discussed above, we have

$$x_{i,A} = R_{iA} + R_{iB}u_{B,A}.$$

Thus $\underline{F} = \underline{R}$ to lowest order. Since $\underline{R} = \underline{R}^{-1t}$, we could equally well use $\underline{F}^{-1t}$ as a lowest-order approximation to $\underline{R}$. In cartesian coordinates, the components of $\underline{F}^{-1t}$ are $X_{A,i}(\partial X_A/\partial x_i)$. Thus, correct to lowest order, we can express the rotated stress in terms of the unrotated stress in several different-looking forms:

$$\sigma_{ij} = x_{i,A}x_{j,B}\sigma^*_{AB} = X_{A,i}X_{B,j}\sigma^*_{AB} = x_{i,A}X_{B,j}\sigma^*_{AB} = \text{etc.}$$

These expressions differ from one another in their second-order terms, and there is no reason to suppose that any of them are correct to second order, since second-order terms are already neglected in the stress-strain relation for infinitesimal deformations.

The distortion of a volume element can be specified by giving the change of length of every fiber element in it. The deformed length $|\underline{dx}|$ of an element with initial span $\underline{dX}$ can be found from

$$\underline{dx}\cdot\underline{dx} = (\underline{F}\underline{dX})\cdot(\underline{F}\underline{dX}) = \underline{dX}\cdot\underline{G}\underline{dX},$$

where the <u>strain</u> $\underline{G}$ is

$$\underline{G} = \underline{F}^t\underline{F}.$$

In cartesian coordinates, the strain components are

$$G_{AB} = x_{i,A}x_{i,B}.$$

If the deformation is mainly a large rotation, so that $\underline{F}$ is almost orthogonal, then $\underline{G}$ is nearly an identity matrix $\underline{I}$. The small difference between $\underline{G}$ and $\underline{I}$ is a measure of the infinitesimal distortion of the element. In terms of a small displacement $\underline{u}$ and large rotation $\underline{R}$, $\underline{G}$ takes the form

$$G_{AB} = \delta_{AB} + u_{A,B} + u_{B,A},$$

neglecting terms of second order in the displacement gradients. Of course, the rotation $\underline{R}$ cancels out when $\underline{G}$ is evaluated.

In the expression for the un-rotated stress σ^*_{AB} we can replace the change of $u_{C,D}$ by a symmetrized version, $\frac{1}{2}d(u_{C,D} + u_{D,C})$, or thus by $\frac{1}{2}dG_{CD}$. The latter form gives what we are looking for, since $\underline{G}$ can be evaluated without bothering to determine $\underline{u}$. Thus, we obtain

$$\sigma_{ij} = x_{i,A}x_{j,B}c_{ABCD}*d(\frac{1}{2}G_{CD}).$$

Written out in a less abbreviated form, this is

$$\sigma_{ij}(\underline{X},t) = x_{i,A}(\underline{X},t)x_{j,B}(\underline{X},t) \int_{-\infty}^{t} c_{ABCD}(t-t')\frac{1}{2}dG_{CD}(\underline{X},t').$$

Notice that all quantities are evaluated at the same particle $\underline{X}$, and thus not necessarily at a fixed point $\underline{x}$ in space. The deformation gradients preceding the integral, whose function is to account for the rotation, are of course evaluated at the time at which the stress is being evaluated. The past history of rotation does not affect the stress.

Problem: Show that the quantity $-X_{A,i}X_{B,i}$ serves the same purpose as G_{AB} as a measure of small distortions.

4. Relative Strain.

It is sometimes convenient to use the present position $\underline{x}(t)$ (say) of a particle as its label. Let $\underline{x}(t')$ be the position of the same particle at time t'. The relative deformation gradient $\underline{F}(t',t)$ is defined by $d\underline{x}(t') = \underline{F}(t',t)d\underline{x}(t)$, so that in cartesian coordinates,

$$F_{ij}(t',t) = \partial x_i(t')/\partial x_j(t).$$

We see that $\underline{F}(t,t) = \underline{I}$,

$$\underline{F}(t'',t')\underline{F}(t',t) = \underline{F}(t'',t),$$

and thus

$$\underline{F}^{-1}(t,t') = \underline{F}(t',t).$$

The relative strain, the strain at time t' relative to the state at time t, is

$$\underline{G}(t',t) = \underline{F}^{t}(t',t)\underline{F}(t',t).$$

Under a change of reference time, relative strains change in a messy way. For example,

$$\begin{aligned}\underline{G}(t',0) &= [\underline{F}(t',t)\underline{F}(t,0)]^{t}\underline{F}(t',t)\underline{F}(t,0)\\ &= \underline{F}^{t}(t,0)\underline{G}(t',t)\underline{F}(t,0).\end{aligned}$$

In the stress-strain relation, let $\underline{G}(\underline{X},t')$ be the strain relative to the state at time zero, i.e. suppose that the material is undeformed up to time zero. Write it in the present notation as $\underline{G}(t',0)$. Then the change in strain, which refers to a change with respect to t', can be written in terms of $\underline{G}(t',t)$ as

$$d\underline{G}(t',0) = \underline{F}^{t}(t,0)d\underline{G}(t',t)\underline{F}(t,0).$$

If we hide the rotation matrices away by counting them in with the relaxation moduli, i.e. defining

$$c_{ijkl}(t,t-t') = R_{iA}(t)R_{jB}(t)R_{kC}(t)R_{lD}(t)c_{ABCD}(t-t'),$$

then the stress-strain relation takes the form

$$\sigma_{ij}(t) = \int_{-\infty}^{t} c_{ijkl}(t,t-t')\tfrac{1}{2}dG_{kl}(t',t).$$

Here we use either $\underline{F}(t,0)$ or $\underline{F}^{-1t}(t,0)$ as $\underline{R}(t)$, as convenience dictates.

5. Isotropic Materials.

If the relaxation moduli c_{ABCD} are those for an isotropic material, then all of the rotation matrices in the definition of c_{ijkl} cancel out since the modulus tensor must be form-invariant under the full orthogonal group.

(Problem: Verify this by using the explicit expression for c_{ABCD} for isotropic materials).

We obtain

$$\sigma_{ij}(t) = \frac{1}{2}\int_{-\infty}^{t} [\lambda(t-t')\delta_{ij}dG_{kk}(t',t) + 2\mu(t-t')dG_{ij}(t',t)].$$

We should emphasize that any number of other forms of the relation are equally accurate, i.e. correct to first order. For example, we can write

$$\sigma_{ij}(t) = \tfrac{1}{2}x_{i,A}(t)x_{j,B}(t)\int_{-\infty}^{t} [\lambda(t-t')\delta_{AB}dG_{CC}(t') + 2\mu(t-t')dG_{AB}(t')]$$

giving the stress in terms of deformation gradients and strains measured relative to a fixed time.

Problem: Show that the compliance form of the stress-deformation relation expresses $\underline{G}(t,0)$ (say) in terms of the history of the unrotated stress, $X_{A,i}X_{B,j}\sigma_{ij}$

(or $x_{i,A}\, x_{j,B}\, \sigma_{ij}$, etc.). Knowledge of the stress history alone, with no knowledge of the rotation history, is not sufficient to determine the deformation. Give an example to explain why this is true.

6. Fluids.

In problems involving viscoelastic fluids we usually simplify matters by supposing that the fluid is incompressible. This is a reasonable idealization in most problems, since the fluid under consideration probably has a compressibility about like that of water. The term involving the Lamé modulus $\lambda(t)$ in the stress-strain relation is replaced by an arbitrary reaction pressure:

$$\sigma_{ij}(t) = -p(t)\delta_{ij} + \int_{-\infty}^{t} \mu(t-t')dG_{ij}(t',t).$$

The relative strain $\underline{G}(t',t)$ is especially useful in fluid dynamical problems. Since the shearing stress relaxation modulus $\mu(t)$ is relatively small when t exceeds two or three mean relaxation times T, the material remembers only the most recent part of the strain history. The strain measured with respect to some fixed reference state is not especially relevant.

Ordinarily, the strain $\underline{G}$ should be close to the identity $\underline{I}$ at <u>all</u> times in order for distortions to be small enough to justify using the linear stress-relaxation modulus. However, for fluids, if the material suffered a large distortion sometime in the distant past, what does it matter? If enough time has elapsed for the stress due to this large distortion to have relaxed almost completely to zero, then it is not terribly important that it be evaluated absolutely correctly. If we evaluate it by using the linear stress relaxation modulus, no harm will be done. It <u>is</u> important that the relative strain $\underline{G}(t',t)$ be close to $\underline{I}$ when $t-t'$ is smaller than $2T$ or $3T$, say; if there has been a large deformation recently, the accompanying stress may still be quite considerable, and we have no reason to believe that it is given correctly by the linear stress-relaxation modulus.

7. Example: Steady Simple Shearing.

We have talked about steady shearing motions much earlier, without needing to comment on the fact that the deformation grows infinitely large in such a motion. As an exercise, let's look at what the present formulation has to say about such motions.

Suppose that each particle moves in the x_1 direction with a speed γx_2, the shear rate γ being constant. The position of a particle at time t' is related to its position at time t by

$$\underline{x}(t') = \underline{x}(t) - (t-t')\gamma x_2 \underline{i}.$$

Consequently, the relative deformation gradient is

$$\underline{F}(t',t) = \underline{I} - \gamma(t-t')\underline{i}\ \underline{j},$$

and the relative strain is

$$\begin{aligned}\underline{G}(t',t) &= [\underline{I} - (t-t')\gamma\underline{j}\ \underline{i}]\cdot[\underline{I} - (t-t')\gamma\underline{i}\ \underline{j}] \\ &= \underline{I} - (t-t')\gamma(\underline{ji} + \underline{ij}) + (t-t')^2\gamma^2\underline{jj}.\end{aligned}$$

Thus $\underline{G}$ is close to $\underline{I}$ over a time of the order of T if γT is small. We are justified in using the linear relaxation modulus only in such cases.

Earlier we have mentioned "second-order" terms without saying explicitly what is meant by that. In the present problem we see that γT is the parameter whose smallness is required. We see that the strain history involves a term proportional to $(\gamma T)^2$. In the present case, this is the "second-order" term, and it should be ignored. The stress will indeed involve contributions proportional to $(\gamma T)^2$, but there is no reason to suppose that the constitutive equation used here gives them correctly.

Neglecting the non-linear term in the strain history, then, the change of

$\underline{G}$ with respect to t' is

$$d\underline{G}(t',t) = \gamma(\underline{ij} + \underline{ji})dt',$$

and the stress-deformation relation gives

$$\underline{\sigma}(t) = \mu*d\underline{G} = \gamma(\underline{ij} + \underline{ji}) \int_0^\infty G(t')dt' = \eta_0(\underline{ij} + \underline{ji}).$$

Thus, with considerable awkwardness, we find what we knew at the outset: in the linear approximation, all stress components are zero except σ_{12} $(= \sigma_{21})$, which is equal to η_0. (There can also be an isotropic pressure $p(t)$ which was ignored above.)

8. Example: Oscillatory Shearing.

As another exercise, consider an oscillatory shearing motion, in which the position of a particle is given as a function of time by

$$\underline{x}(\underline{X},t) = \underline{X} + \underline{i}A\sin(\omega t)X_2.$$

Then

$$\underline{F}(t,0) = \underline{I} + A\sin(\omega t)\underline{i}\ \underline{j},$$

and

$$\underline{G}(t,0) = \underline{I} + A\sin(\omega t)(\underline{ij} + \underline{ji}) + A^2\sin^2(\omega t)\underline{jj}.$$

Evidently $\underline{G}$ is close to $\underline{I}$ at all times if A is small. In that case the term involving A^2 should be neglected. Moreover, in

$$\underline{\sigma}(t) = \underline{F}(t,0)(\mu*d\underline{G})\underline{F}^t(t,0),$$

(leaving aside an arbitrary isotropic pressure) where the deformation gradients appear in order to account for rotations, we should use $\underline{F} = \underline{I}$ since otherwise spurious terms involving higher powers of A will appear. Consequently, we obtain

$$\underline{\sigma}(t) = A\omega(\underline{ij} + \underline{ji})\int_{-\infty}^{t} \mu(t-t')\cos(\omega t')dt',$$

the expected result.

Notice that the strain rate, whose amplitude is $A\omega$, can be arbitrarily large. Regardless of how large the frequency may be, the distortion of each material element is small at all times, i.e. $\underline{G} \cong \underline{I}$, if the strain amplitude A is small. Of course, $A\omega$ is not dimensionless, so there is no meaning to a statement that $A\omega$ is large anyway.

Problem: Work the preceding exercise over again, using the formulation in terms of relative strain, $\underline{G}(t',t)$.

9. Motions with Uniform Velocity Gradient.

Let us consider what would appear to be an exceptionally dull class of problems: We suppose that an infinite body of incompressible fluid is undergoing motion with a velocity gradient constant in space and time. With velocity $\underline{v} = \underline{N}\underline{x}$, where $\underline{N}$ is a constant matrix, we must first find the deformation gradient history. Now,

$$\frac{\partial F_{ij}(t',t)}{\partial t'} = \frac{\partial}{\partial t'}\frac{\partial x_i(t')}{\partial x_j(t)} = \frac{\partial}{\partial x_j(t)}\frac{\partial x_i(t')}{\partial t'} = \frac{\partial v_i(t')}{\partial x_j(t)},$$

and thus

$$\frac{\partial F_{ij}(t',t)}{\partial t'} = \frac{\partial v_i(t')}{\partial x_n(t')}\frac{\partial x_n(t')}{\partial x_j(t)} = N_{ik}F_{kj}(t',t).$$

Reverting to direct notation, we have

$$\dot{\underline{F}}(t',t) = \underline{N}\underline{F}(t',t).$$

The dot means differentiation with respect to t'. With $\underline{F}(t,t) = \underline{I}$, we obtain

$$\underline{F}(t',t) = \exp[(t'-t)\underline{N}],$$

where the matrix exponential is defined by

$$\exp(\underline{N}t) = \sum \frac{1}{n!} t^n \underline{N}^n.$$

Problems:

1. Show that $\exp \underline{N} \exp \underline{M} = \exp(\underline{N}+\underline{M})$ only if $\underline{NM} = \underline{MN}$.

2. Verify that $\underline{F}(t'',t')\underline{F}(t',t) = \underline{F}(t'',t)$ is satisfied by the matrix exponential form of $\underline{F}$ given above.

To simplify matters, let us restrict attention to plane flows, in which $\underline{N}$ is essentially a 2×2 matrix (i.e. $N_{3i} = N_{i3} = 0$). With no loss of generality we can take the coordinate axes in the directions of the principal axes of the symmetric part of $\underline{N}$, so that $\underline{N}$ takes the form

$$\underline{N} = \epsilon(\underline{ii} - \underline{jj}) + \omega(\underline{ji} - \underline{ij}).$$

The two diagonal components ϵ and $-\epsilon$ must cancel out because the flow is isochoric.

What we are trying to do is to find out what in the world $\exp(\underline{N}t)$ means. To do this, from the definition, we need to evaluate all of the powers of $\underline{N}$. We find first that

$$\underline{N}^2 = -(\omega^2 - \epsilon^2)(\underline{ii} + \underline{jj}) = -\Omega^2\underline{\Delta},$$

where $\Omega^2 = \omega^2-\epsilon^2$ and $\underline{\Delta}$ is a two-dimensional version of a unit matrix. From this it follows that

$$\underline{N}^{2n} = (-1)^n \Omega^{2n} \underline{\Delta} \quad (n \neq 0), \quad \underline{N}^{2n+1} = (-1)^n \Omega^{2n+1} (\underline{N}/\Omega).$$

Or course, $\underline{N}^0 = \underline{I} = \underline{\Delta} + \underline{kk}$.

There is now no trouble in summing the series that defines $\underline{F}(t',t)$, and we obtain

$$\underline{F}(t',t) = \underline{kk} + \underline{\Delta} \cos \Omega(t'-t) + (\underline{N}/\Omega) \sin \Omega(t'-t).$$

Then the relative strain is

$$\underline{G}(t',t) = \underline{kk} + \underline{\Delta} \cos^2 \Omega(t'-t) + (\underline{N}^t \underline{N}/\Omega^2) \sin^2 \Omega(t'-t)$$
$$+ \Omega^{-1} (\underline{N} + \underline{N}^t) \sin \Omega(t'-t) \cos \Omega(t'-t).$$

The notation has been set up as if ω were bigger than ϵ, so that Ω is real. The other way around, sin and cos would be replaced by sinh and cosh to keep everything in real form.

Since $\underline{G}$ has to do with changes of fiber lengths, we see that Ω has the significance of a frequency of oscillation of the length of a fiber. When ω is greater than ϵ, so that Ω is real, the fluid mainly rotates. Particles travel in elliptical orbits. A fiber that is radial at any one instant is always radial. With one end at the origin and the other moving around an ellipse, the fiber alternately stretches and contracts, from the length of the minor axis to that of the major axis. Actually there are two cycles of stretching in each complete rotation, so the frequency of oscillation of the fiber length is 2Ω.

Since

$$(\underline{N} + \underline{N}^t)/\Omega = 2\epsilon(\underline{ii} - \underline{jj})/(\omega^2 - \epsilon^2)^{1/2}$$

and

$$\underline{N}^t\underline{N}/\Omega^2 = [(\omega^2+\epsilon^2)\underline{\Delta} - 2\epsilon\omega(\underline{ij} + \underline{ji})]/(\omega^2-\epsilon^2),$$

we see that $\underline{G}$ is close to $\underline{I}$ at all times, and thus distortions are always small, if ϵ/ω is small. In that case, terms of order $(\epsilon/\omega)^2$ should be neglected in evaluating the stress to first order. With neglect of such terms, $\Omega = \omega$ and for $\underline{G}$ we obtain

$$\underline{G}(t',t) = \underline{I} - (\epsilon/\omega)(\underline{ij} + \underline{ji})[1 - \cos 2\omega(t'-t)]$$
$$+ (\epsilon/\omega)(\underline{ii} - \underline{jj})\sin 2\omega(t'-t).$$

Then

$$\frac{d}{dt'}G(t',t) = 2\epsilon(\underline{ij} + \underline{ji})\sin 2\omega(t-t')$$
$$+ 2\epsilon(\underline{ii} - \underline{jj})\cos 2\omega(t-t').$$

Let us write the complex viscosity $G^*(\omega)/i\omega$ as $\eta_1 - i\eta_2$. Then the stress is found to be

$$\underline{\sigma}(t) = -p(t)\underline{I} + 2\epsilon(\underline{ij} + \underline{ji})\eta_2(2\omega) + 2\epsilon(\underline{ii} - \underline{jj})\eta_1(2\omega).$$

Anyone knows that the stress can't possibly depend upon the vorticity. Nevertheless, our result shows that 2ω is mainly what the stress does depend upon, and for a good and understandable reason. As the fluid flows in almost circular elliptical paths, each fluid element is jiggled at a radian frequency equal to the vorticity, so the stress response is governed by the dynamic viscosity at that frequency. The response is nothing like linear in its dependence on ω, either. Although we are dealing with a constitutive equation that is something like linear, at least using only the linear response function, we find that the stress depends on the motion in a remarkably non-linear way.

Notice that the validity of the result does not depend on whether the strain-rate ϵ is small or not. It can be arbitrarily large, provided that ω is much larger still.

Problems:

1. Show that it is possible to determine $p(\underline{x},t)$ in such a way that the preceding motion satisfies the momentum equation.

2. Show that if $\underline{N}^t = -\underline{N}$, then $\exp(\underline{N}t)$ is orthogonal.

3. What sort of motion is there when $\epsilon = \omega$? When $\epsilon > \omega$? When $\omega = 0$?

Problem: The Maxwell-Chartoff Rheometer.

Consider an apparatus consisting mainly of two parallel rotating discs, rotating with equal angular velocities ω, with axes that are parallel but not coincident. One disc is in the plane $y = 0$, and it rotates about the y-axis. The other is in the plane $y = h$, rotating about the axis $x = 0$, $z = -k$. There is a blob of viscoelastic material between the two plates. Because of the curious arrangement, it is kneaded as the plates turn. If the plates are close enough together we can mainly ignore the momentum equation and claim that the obvious kinematics gives the right answer. (Why is that? It can't always be right. What is being assumed?) The simplest idea is that the layer of fluid on a plane y = constant rotates about the point $x = 0$, $z = -ky/h$. What is the velocity field, under that assumption? Show that the velocity gradient is independent of position and time. Find the relative deformation gradient history. Find the strain history. What parameter should be small in order that there never be much distortion of any material element? Can you answer this without any calculation? Evaluate the stress components, and in particular show that they are independent of position and time. It is surely true that energy is being dissipated as the blob is kneaded. Where is the power coming from? What forces are acting through a distance? Suppose that the material is a Newtonian (Navier-Stokes) viscous fluid. Can you adjust the preceding kinematical assumptions a little so that is it possible to satisfy the momentum equation exactly? Whether you can or not, hunt down Walters' paper on the subject.

10. Anisotropic Fluids.

Let us briefly consider a special kind of anisotropic fluid, a liquid crystal with a needle-like structure. Suppose that in some reference state the needles are lined up in the X_3 direction. Let us assume that the material is transversely isotropic about the preferred direction, so that the modulus tensor c_{ABCD} is invariant under rotations about that direction and under reflections in planes parallel or perpendicular to the X_3 direction. All tensors (of any rank) with such invariance are linear combinations of outer products of the Kronecker delta δ_{AB} and the projection operator $\delta_{3A}\delta_{3B}$. Since c_{ABCD} is symmetric on AB and also on CD, by enumerating the possibilities we find that c_{ABCD} must be a linear combination of the tensors

$$\delta_{AB}\delta_{CD},\ \delta_{AC}\delta_{BD} + \delta_{AD}\delta_{BC},\ \delta_{AB}\delta_{3C}\delta_{3D},\ \delta_{3A}\delta_{3B}\delta_{CD},$$

$$\delta_{AC}\delta_{3B}\delta_{3D} + \delta_{BC}\delta_{3A}\delta_{3D} + \delta_{AD}\delta_{3B}\delta_{3C} + \delta_{BD}\delta_{3A}\delta_{3C},$$

$$\delta_{3A}\delta_{3B}\delta_{3C}\delta_{3D}.$$

The time-dependent coefficients of these tensors are the six independent relaxation moduli allowed by symmetry. To simplify matters a little, let us suppose that the material is incompressible. Then since the first and third terms above would make a contribution to σ^*_{AB} proportional to δ_{AB}, which would be indistinguishable from the arbitrary reaction pressure, we can ignore those terms. Moreover, in the product $c_{ABCD}u_{C,D}$, the fourth term on the list would make a contribution involving $u_{C,C}$, which is zero (or rather, second-order) if the motion is isochoric. Hence we can omit that term as well.

Now, recall the device, which at the time seemed only a notational trick, of defining a set of rotated moduli $c_{ijkl}(t,t-t')$ by operating on the response functions $c_{ABCD}(t-t')$ with the rotation matrices $R_{iA}(t)$ defined by the deformation. If we transform each of the three remaining tensors on the list above in this way, we obtain

$$\delta_{ik}\delta_{jl} + \delta_{il}\delta_{jk}, \quad a_i a_j a_k a_l, \quad \text{and}$$

$$\delta_{ik}a_j a_l + \delta_{jk}a_i a_l + \delta_{il}a_j a_k + \delta_{jl}a_i a_k,$$

where the unit vector $\underline{a}(t)$ is in the direction of the needles at the particle considered, taking into account their possibly large rotation:

$$a_i(t) = R_{i3}(t).$$

Recall the form of the stress-deformation relation. With the reaction pressure thrown in, it is

$$\sigma_{ij}(t) = -p(t)\delta_{ij} + \frac{1}{2}\int_{-\infty}^{t} c_{ijkl}(t,t-t')dG_{kl}(t',t)$$

With the form of c_{ijkl} as worked out, we obtain

$$\sigma_{ij}(t) = -p(t)\delta_{ij} + \int_{-\infty}^{t} \mu_1(t-t')dG_{ij}(t',t)$$
$$+ a_i(t)a_j(t) \int_{-\infty}^{t} \mu_2(t-t')a_k(t)dG_{kl}(t',t)a_l(t)$$
$$+ \int_{-\infty}^{t} \mu_3(t-t')[a_i(t)dG_{jk}(t',t) + a_j(t)dG_{ik}(t',t)]a_k(t).$$

Needless to say, this is much too complicated for anyone to have done anything with it. To obtain some results that might actually improve our understanding of the mechanical behavior of liquid crystals, in our present state of relative ignorance it would be better to try a less detailed model. For example, each integral might better be approximated by a viscosity coefficient (equal to the integral of the modulus) times the present value of the strain rate.

Although there is a large body of literature on liquid crystals, dating back a very long time, most of the work on the theory of their mechanical behavior is fairly recent. See the review article by J. L. Ericksen (Applied Mechanics Reviews, November, 1967) for a start.

CHAPTER VIII

SLOW VISCOELASTIC FLOW

Constitutive equations based on the linear response functions are not adequate for most problems of viscoelastic flow. In the present chapter we first discuss the general problem of choosing a constitutive equation that can describe the stress response adequately in whatever specific problem is at hand. We then turn to a specific class of flows for which the Navier-Stokes equation is a good first approximation. These flows are slow in the sense that nothing changes much in one relaxation time, but may be arbitrarily fast if by "fast" one means something to do with the Reynolds number. We discuss the nature of approximations based on this notion of slowness and show how to solve flow problems within this approximation scheme.

1. Viscoelastic Flow.

In the case of simple shearing at a constant rate, we found that the stress can be evaluated by using the linear stress-relaxation modulus only if γT is small. In this limit, the shearing stress is proportional to the shear rate, $\sigma = \eta_0 \gamma$. Actually, when a fluid is noticeably viscoelastic at all, the stress is always markedly non-linear in its dependence on the shear rate: $\sigma = \eta(\gamma^2)\gamma$. The ratio σ/γ is called the apparent viscosity. The apparent viscosity usually decreases sharply as the shear rate increases (for polymer melts and solutions) and it is often rather difficult to shear the fluid so slowly that the apparent viscosity is close to its limiting value at zero shear-rate, $\eta(0) = \eta_0$.

The materials that we are likely to recognize as fluids are very easy to shear. With forces that are easily accessible to the experimentalist, very large deformations can be produced. If the mean relaxation time T is something on a human scale, say one second, then it may be possible to produce a very large deformation within one relaxation time. The stress understandably depends non-linearly on the strain when the amount of strain that the material can remember is large.

People have a very natural and understandable desire to find the stress-deformation relation that describes the behavior of viscoelastic fluids. This search for the philosopher's stone has produced a great deal of very valuable work, and I imagine that it will continue to do so in the future. Furthermore, I am optimistic enough to hope that someday, for at least one specific viscoelastic fluid, we may find a reasonably simple and specific expression for the stress as a functional of the history of deformation.

Nevertheless, if we are to deal with actual knowledge rather than interesting speculations, at the moment (and probably forever) we must use constitutive equations whose forms are adapted to the conditions of flow under consideration. We do not specify one equation that describes all of the mechanical behavior of some particular material. Rather, we use constitutive equations of a variety of different forms, the form being governed more by the kind of flow considered than by detailed material properties.

2. Flow Diagnosis.

The question of what equation to use in solving a given problem is not a matter of settled opinion, because we are relatively ignorant both about material properties and about efficient ways of describing them mathematically. However, there are some limiting cases in which it is fairly certain which equation should be used.

Laminar shearing flows can be classified, loosely, by assigning to each flow a characteristic shear amplitude A and a frequency ω. We will interpret these parameters very broadly, but for present purposes, to fix ideas one may think of ordinary sinusoidal shearing of a thin layer of fluid. To characterize the fluid in the least complicated way possible, while at the same time bringing in a property that is distinctly viscoelastic rather than viscous or elastic, we use the mean relaxation time T or some estimate of it. We combine the dimensional parameters ω and T as a dimensionless frequency ωT. The shear amplitude A is dimensionless as it stands.

We plot ωT and A on distorted scales so that all values from zero to infinity lie in a square. Each point of the square represents some particular flow or class of flows. In effect, the whole infinite-dimensional space of shearing strain histories is projected onto two dimensions. The hope is that the two dimensions chosen are the two most important.

∞

A

Viscometric Flow: $\sigma = \eta(\gamma^2)$

Finite Elasticity: $\sigma = \kappa\mu(\kappa^2)$

?

Navier-Stokes:

$$\sigma = \eta(0)\gamma$$
$$= \gamma \int_0^\infty G(t)dt$$

Linear Viscoelasticity:

$$\sigma(t) = \int_{-\infty}^{t} G(t-t')d\kappa(t')$$

Infinitesimal Elasticity:

$$\sigma = \mu(0)\kappa$$
$$= G(0+)\kappa$$

0

∞

ωT

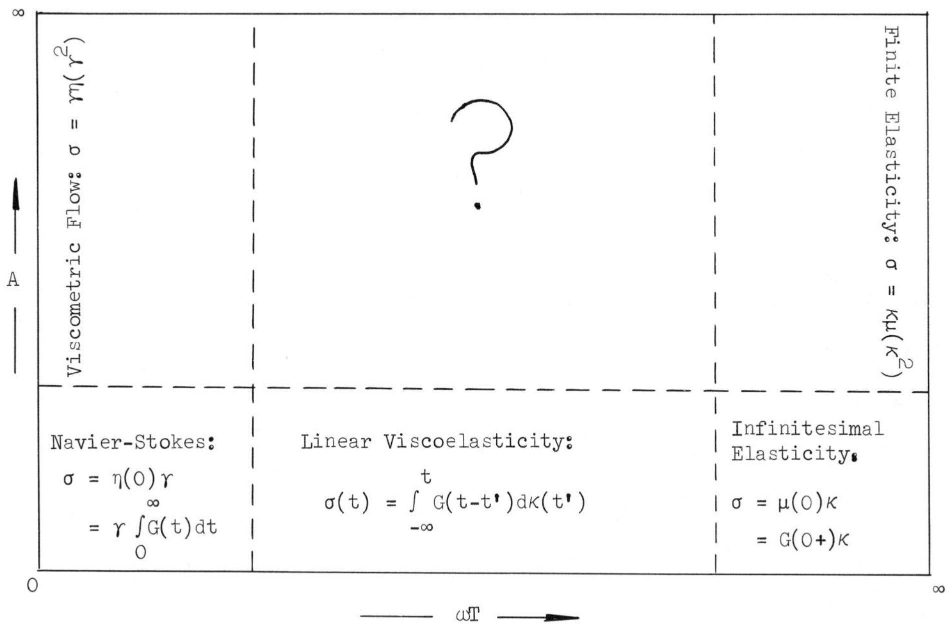

Near the edge $A = 0$ we are dealing with the infinitesimal-amplitude motions for which linear viscoelasticity theory is valid. We would wish to include motions with small distortions but large displacements and rotations in this class, of course.

The edge $\omega T = \infty$ corresponds to finite elastic deformations. As we have seen in linear viscoelasticity theory, the response of the material in the limit of infinite frequency or zero time is the instantaneous elastic response. If the total duration of an experiment is t_o, we might interpret the characteristic frequency ω as $1/t_o$. Then saying that ωT is large means that the time of observation is short in comparison to the mean relaxation time. If not enough time has elapsed for there to be any noticeable stress-relaxation, we call the response elastic. Along this edge, the shear amplitude A would stand for a typical amount of shear, as estimated from the boundary conditions of the specific problem considered. In one-dimensional form, the constitutive equation describing behavior in this region is simply an expression for the shearing stress as some non-linear function of the concurrent value of the amount of shear, $\sigma = \mu(\kappa^2)\kappa$, say.

At the corner $\omega T = \infty$, $A = 0$, we have, simultaneously, the short-time limit of linear viscoelasticity theory and the small-deformation limit of elasticity theory. From the point of view of viscoelasticity theory we would write the stress-strain relation as $\sigma = G_g\kappa$, and as a limit of finite elasticity theory we would write it as $\sigma = \mu(0)\kappa$, where $\mu(\kappa^2)$ is the apparent shear modulus. Comparison shows that $G_g = \mu(0)$. Whenever we find a region in which two different forms of constitutive equations are valid simultaneously, we can find relations between their parameters in this way.

In terms of sinusoidal shearing, the limit of zero frequency is the case of steady shearing. In this limit the shearing stress is some non-linear function of the shear rate, $\sigma = \gamma\eta(\gamma^2)$. Along the edge $\omega T = 0$, we interpret the characteristic shear amplitude A as γT, the amount of shear in one relaxation time, since only the part of the deformation that the fluid can remember is relevant.

At the corner $A = \omega T = 0$ we have steady shearing in the limit of zero shear rate, $\sigma = \eta(0)\gamma$, and infinitesimal viscoelasticity in the limit of steady shearing $(\kappa = \gamma t)$. Because of this overlap, we obtain the result that the apparent viscosity at zero shear-rate, η_o, is the integral of the stress-relaxation modulus. This region, in which the stress depends linearly on the concurrent value of the strain-rate, is the realm of Newtonian (Navier-Stokes) fluid dynamics. (It may seem strange to call classical fluid dynamics a low shear-rate approximation; remember that by "low" we mean small in comparison to $1/T$.)

Nothing very systematic is known about the interior region in which A is large and ωT is neither large nor small. Here we are dealing with large deformations, with a time scale so commensurable with that defined by the material that it is not possible to overlook viscoelastic effects. One way to envisage the kind of equation that would describe behavior in this region is to think of the stress as an analytic function of the strain increments $d\kappa$. By expanding the stress in powers of these variables, we obtain

$$\sigma(t) = \int_{-\infty}^{t} G(t-t')d\kappa(t') + \int_{-\infty}^{t}\int_{-\infty}^{t} G_2(t-t',t-t'')d\kappa(t')d\kappa(t'') + \ldots \ .$$

The linear terms of the expansion give the familiar linear stress-relaxation integral. The quadratic terms give rise to a double integral, and so on. Actually, in the case of shearing response the stress must be an odd function of the shear history, so the coefficients of the quadratic terms must vanish, $G_2 = 0$. The leading non-linear contribution then comes from a triple integral.

It would require an enormous number of experiments to determine the modulus in the triple integral, a function of three variables, to anything like the same accuracy as the linear stress-relaxation modulus. Even having done it, we would still have a relation valid only for small strains. Consequently, this multiple-integral type of expansion has not been used very much. It is more useful as a conceptual device than as a practical way to organize data and solve problems.

Problem: What equations apply to the edges $\omega T = \infty$, $A = 0$, and $\omega T = 0$ if the material is a solid?

3. Slow Viscoelastic Flow: Asymptotic Approximations.

Any of the constitutive equations that describe behavior around the edges of the flow diagnosis diagram can be regarded as the lowest-order term in some kind of series expansion that, we hope, might describe behavior at points more toward the interior. For example, we have just seen how the linear viscoelastic constitutive equation can be regarded as the first term of an expansion in terms of multiple integrals. Let us examine a simpler kind of expansion, covering a more limited region, the neighborhood of the Navier-Stokes corner.

To understand the motivation for the kind of expansion to be postulated, first consider only the one-dimensional relation between shearing stress and shearing strain, as a model. Write the linear viscoelastic relation as

$$\sigma(t) = \int_{0}^{\infty} G(t')\kappa'(t-t')dt' .$$

Now, suppose that the shear rate (or rather $\kappa' T$) is small (so that we can use the linear viscoelastic relation at all) and that it is changing only slowly. If the modulus $G(t')$ becomes negligibly small within a time of order T ($2T$, or $10T$, say) then to evaluate $\sigma(t)$ we need to know the history $\kappa'(t-t')$ only during this most recent interval of order T. Let us suppose that the shear rate is varying so smoothly that during this interval, it can be represented accurately by a few terms of its Taylor series expansion:

$$\kappa'(t-t') = \kappa'(t) - \kappa''(t)t' + \tfrac{1}{2}\kappa'''(t)(t')^2 - \dots .$$

By using this expansion in the integral and carrying out the integration, we obtain

$$\sigma(t) = \eta_0 \kappa'(t) - \eta_0 T \kappa''(t) + \dots .$$

The higher-order terms of the series involve higher derivatives of κ, evaluated at time t, and their coefficients are the higher moments of the modulus, $\int_0^\infty G(t)t^n dt$.

It would appear that if the shear rate is fairly constant over intervals of order T, it might be sufficiently accurate to use only the leading term and neglect the rest. Thus, for motions that are sufficiently slow (with T as time unit) we obtain a linear relation between the stress and the concurrent value of the strain-rate, or, in a three-dimensional version, the Navier-Stokes equation. It would appear that if the shear rate is a little more variable, we might need to include the term proportional to the second derivative as well.

Appearances can be deceiving, however. The kind of series that we have obtained for the stress is likely to be an asymptotic expansion rather than a convergent series. Because of the perverse way in which asymptotic expansions behave, we do not necessarily get better accuracy by keeping more terms of the series. It is truer to say that if the first term is very accurate by itself, we can obtain a more refined result by keeping the second term as well (which is necessarily small in comparison to the first term in such cases) but if the first term is not a good approximation to the stress, it will only make matters worse to include the second term, because we are off on the wrong foot.

<u>Example</u>: Consider a Maxwell model, $G(t) = G_g \exp(-t/T)H(t)$. Consider simple shearing starting from rest, $\kappa(t) = \gamma t H(t)$. For $t > 0$, the expansion of $\kappa'(t-t')$ in powers of t' is simply the one term γ. The series for the stress is then also a single term, $\eta_o \gamma$. This is accurate when enough time has elapsed that the break at time zero has been forgotten, but of course it is nonsense for t of the order of T or smaller. In this example the power series $\kappa'(t-t') = \gamma$ is convergent for all t', trivially, but the series is actually equal to $\kappa'(t-t')$ only for $t' < t$. We make the error by pretending that the series represents the shear rate for all time.

<u>Problem</u>: Evaluate the linear stress relaxation integral exactly, and show how the break at time zero is forgotten when t grows large in comparison to T.

<u>Example</u>: Using a Maxwell model again, consider the shearing history $\kappa(t) = AH(t)\log(1+t)$. For $t > 0$, the expansion of $\kappa'(t-t')$ is

$$\kappa'(t-t') = \frac{A}{1+t} \sum \left(\frac{t'}{1+t}\right)^n .$$

We pretend that this converges for all t', which it does not, and that it is equal to $\kappa'(t-t')$ for $0 \leq t' < \infty$, which is not true. By using it in the stress relaxation integral we obtain a series representation for the stress,

$$\sigma(t) = \frac{A\eta_o}{1+t} \sum n! \left(\frac{T}{1+t}\right)^n .$$

This new series does not converge at all, but that is no objection. For large enough t, each term of the series is smaller than the one before it, up to some term, after which the terms start growing larger again. By summing up to the smallest term and forgetting the rest we can evaluate the stress very accurately. Provided that $t \gg T$, so that the break at time zero has been forgotten, the error is of the order of the first term omitted.

Problem: By expressing $\kappa'(t-t')$ as a finite sum plus remainder, and using the correct limits of integration, verify the preceding statements.

4. Slow Viscoelastic Flow: Three-Dimensional Equations.

Suppose that we have a problem in which U is a characteristic velocity and L is a typical length, so that U/L is a typical shear rate. Then the amount of shear in one relaxation time is $A = TU/L$. To estimate a frequency ω, suppose that the material derivative of any function is $O(\omega)$ times that function. Suppose further that $d/dt = O(U/L)$, as for a steady flow. Then $\omega T = TU/L$. Let us suppose that TU/L is small. Thus, we are considering flows represented on the flow diagnosis diagram by points near the Navier-Stokes corner, on the diagonal $A = \omega T$.

For the general three-dimensional case, we need to consider the relative strain history $\underline{G}(t-t',t)$. For sufficiently slow, smooth motions, we can represent the strain history over the interval that matters $(O(T))$ by a few terms of its expansion in powers of t',

$$\underline{G}(t-t',t) = \sum \frac{1}{n!} \underline{A}_n(t)(-t')^n.$$

The derivatives $\underline{A}_n(t)$ are called Rivlin-Ericksen tensors:

$$\underline{A}_n(t) = \frac{d^n}{ds^n} \underline{G}(s,t)\Big|_{s=t}.$$

The time derivative is a material derivative, of course; G is understood to be a function of the particle labels as well as of the time arguments shown explicitly.

Since $\underline{A}_n$ is the n-th material derivative of $\underline{G}$, which is dimensionless, then $\underline{A}_n = O(U/L)^n$. In the series expansion of $\underline{G}$, then, the n-th term is $O(TU/L)^n$. This is not true for all values of the time lag t', of course, but it is true for that part of the strain history that the fluid can still remember at time t.

We expect that if TU/L is small then it should be possible to evaluate the stress approximately by pretending that the strain history is equal to the first term of its series expansion, $\underline{G} \cong \underline{A}_o = \underline{I}$. With this approximation, i.e. no motion, we would conclude that the stress must be simply an isotropic pressure, as for a completely relaxed fluid. Although this must be approximately correct, we are, of course, interested in the effect of the motion on the stress. Thus, we consider a more refined approximation, $\underline{G}(t-t',t) = \underline{I} - \underline{A}_1(t)t'$, assuming that $\underline{A}_1 T$ is only a small perturbation. With this approximation, we expect to be able to evaluate the stress by supposing that it is a function of $\underline{A}_1$. $\underline{A}_1$ is (twice) the classical strain-rate tensor, and the resulting approximation for the stress is the Navier-Stokes constitutive equation.

Provided that the Navier-Stokes equation is a more accurate approximation than a pure isotropic pressure, we may be able to evaluate the stress still more accurately by using a three-term approximation to the strain history, $\underline{G} = \underline{I} - \underline{A}_1 t' + \frac{1}{2}\underline{A}_2 t'^2$. In that case, since terms of order $(TU/L)^2$ are being retained in the strain history, we expect to be able to evaluate the stress correct to order $(TU/L)^2$ as well, by using an appropriate function of $\underline{A}_1$ and $\underline{A}_2$. Of course, this expectation may be ill-founded. A given fluid knows nothing of our personal lives, much less our ideas about mathematical convenience. Nevertheless, the model we are engaged in postulating does seem to work pretty well for sufficiently slow, smooth motions, for a great many different fluids.

Let us write the stress as

$$\underline{\sigma} = -p\underline{I} + \underline{S},$$

where p is the reaction pressure and $\underline{S}$ is the part produced by the motion. We suppose that the latter can be represented as a sum,

$$\underline{S} = \underline{S}_1 + \underline{S}_2 + \underline{S}_3 + \dots,$$

where $\underline{S}_n$ is of order $(TU/L)^n$ (times a modulus). We intend to try to approximate

$\underline{S}_n$ by a function of the Rivlin-Ericksen tensors. We need isotropic functions of these tensors, for an isotropic fluid, and we will use analytic functions until or unless the model fails to jibe with experiment. The smooth, isotropic functions of various orders that can be formed from the kinematic tensors are

$$\begin{aligned}
&U/L: && \underline{A}_1, \quad \underline{I}\,\mathrm{tr}\underline{A}_1, \\
&(U/L)^2: && \underline{A}_2, \quad \underline{A}_1^2, \quad \underline{A}_1\mathrm{tr}\underline{A}_1, \quad \underline{I}(\ldots), \\
&(U/L)^3: && \underline{A}_3, \quad \underline{A}_1\underline{A}_2 + \underline{A}_2\underline{A}_1, \quad \underline{A}_2\mathrm{tr}\underline{A}_1, \quad \underline{A}_1\mathrm{tr}\underline{A}_2, \quad \underline{A}_1^2\mathrm{tr}\underline{A}_1, \\
& && \underline{A}_1\mathrm{tr}\underline{A}_1^2, \quad \underline{A}_1(\mathrm{tr}\underline{A}_1)^2, \quad \underline{A}_1^3, \quad \underline{I}(\ldots),
\end{aligned}$$

and so on. Here tr means trace. In writing down this list we use the fact that an isotropic polynomial in the components of a number of tensors can be expressed as a <u>matrix</u> polynomial with $\underline{I}$ as the only possible coefficient matrix. For isotropic materials, you always get a complete list by writing down all of the terms you can think of.

We now suppose that $\underline{S}_n$ is a linear combination of the terms of order $(U/L)^n$, as listed above. Thus, for example,

$$\underline{S}_1 = \frac{1}{2}\lambda_o \underline{I}\,\mathrm{tr}\underline{A}_1 + \eta_o \underline{A}_1.$$

However, we restrict attention to incompressible fluids. The term in $\underline{S}_1$ proportional to $\underline{I}$ makes a contribution to the stress that is indistinguishable from the reaction pressure $-p\underline{I}$, so we could never attach any empirical meaning to that term. Consequently, there is no loss of generality in omitting it. We use

$$\underline{S}_1 = \eta_o \underline{A}_1.$$

In the same way, any contribution to any $\underline{S}_n$ that is proportional to $\underline{I}$ can be omitted, and that is why such terms have not been written out explicitly on the list of terms of various orders.

Incompressibility also implies that $\operatorname{tr}\underline{A}_1 = 0$, so all terms involving $\operatorname{tr}\underline{A}_1$ can be omitted. But there is a much stronger result of this kind. Rivlin has shown that in an isochoric motion, $\operatorname{tr}\underline{A}_n$ can always be expressed identically as an invariant of lower-order kinematic tensors (ZAMP 13, 589(1962)). Thus, any term involving $\operatorname{tr}\underline{A}_n$ can be expressed as a linear combination of other terms on the list, so terms involving $\operatorname{tr}\underline{A}_n$ are redundant.

For $\underline{S}_2$ we use a linear combination of the terms of order $(U/L)^2$, omitting those proportional to $\underline{I}$ or involving $\operatorname{tr}\underline{A}_1$:

$$\underline{S}_2 = -\eta_o T\underline{A}_2 + \eta_o(T+T^*)\underline{A}_1^2.$$

The coefficient of $\underline{A}_2$ must be equal to $-\eta_o T$ for consistency with linear viscoelasticity theory. The coefficient of $\underline{A}_1^2$ is written in a form that is convenient later; the new parameter T^* defined by this coefficient is probably about the same as T or a bit less (empirically).

On the list of terms that can contribute to $\underline{S}_3$, many can be omitted because they involve $\underline{I}$, $\operatorname{tr}\underline{A}_1$, or $\operatorname{tr}\underline{A}_2$. Moreover, $\underline{A}_1^3$ can be expressed as a linear combination of other terms on the list by using the Cayley-Hamilton identity, so it, too, is redundant. In higher orders of approximation such identities would play an even more important role in eliminating redundant terms. The matrix identities that are needed are usually called identities of the Cayley-Hamilton type, although it was Rivlin who invented them. Getting back to business, for $\underline{S}_3$ we use

$$\underline{S}_3 = C_1\underline{A}_3 + C_2\underline{A}_1\operatorname{tr}\underline{A}_1^2 + C_3(\underline{A}_1\underline{A}_2 + \underline{A}_2\underline{A}_1).$$

Here each coefficient has the same dimensions as $\eta_o T^2$, and it would be surprising if any coefficient should be much different from $\eta_o T^2$ in value (apart from sign); certainly, more than an order of magnitude of difference would be an interesting phenomenon requiring explanation.

Notice that $\underline{A}_1\underline{A}_2$ and $\underline{A}_2\underline{A}_1$ must have the same coefficient in order for

$\underline{S}_3$ to be symmetric. Because the stress is symmetric, only symmetrical combinations of the tensors $\underline{A}_n$ need to be considered in any order of approximation.

Problem: Assuming that the stress can also be represented in terms of an expansion in multiple integrals, the terms proportional to $\underline{A}_n$ in the slow-motion expansion must arise from the first integral,

$$\int_0^\infty \mu(t')d\underline{G}(t-t',t).$$

By using the series expansion of $\underline{G}$ here, show that the coefficients of $\underline{A}_1$ and $\underline{A}_2$ (in $\underline{S}_1$ and $\underline{S}_2$, respectively) must be η_0 and $-\eta_0 T$, as asserted. Express the coefficient of $\underline{A}_3$ in the expression for $\underline{S}_3$ in terms of the linear stress-relaxation modulus $\mu(t)$.

5. Orders of Approximation in Slow Motions.

In the preceding discussion we treated $\underline{A}_1$ as a first-order term, and both $\underline{A}_2$ and $\underline{A}_1^2$ as second-order terms. This makes sense in those problems of steady motion in which the two flow diagnosis parameters A and ωT are both equal to TU/L. In unsteady (but smooth) motions, we can say that $\underline{A}_n$ is of order $A\omega^n$ for some appropriate choice of A and ω. Depending on the relative sizes of A and ωT, i.e. depending on just exactly how the analysis is being carried out and what limits are being considered, it may happen that $\underline{A}_1^2$ is a "second-order" term and $\underline{A}_2$ is not, since $\underline{A}_2 = O(A\omega^2)$ and $\underline{A}_1^2 = O(A^2\omega^2)$. Thus, the particular ordering of terms depends on the particular problem; there is nothing universal about the estimates we used.

What stays the same in all slow-motion approximations is the general notion of representing the stress as an ordered sequence of terms, each term being a matrix polynomial involving the Rivlin-Ericksen tensors.

Problem: Consider sinusoidal shearing of a thin layer of fluid, with shear amplitude A and frequency ω. Consider a set of such experiments, all with the same amplitude A but with smaller and smaller values of ω. What is the "lowest-order" approximation to the stress in the limit $\omega T \to 0$? (Consider only shearing stress.)

6. The Rivlin-Ericksen Tensors.

In general, in viscoelastic flow problems one must be concerned with a Lagrangian description of the motion, like it or not, because it is necessary to keep track of particles in order to evaluate their strain histories. The situation is much simpler in cases of slow, smooth motion, because in place of the strain history we can use the Rivlin-Ericksen tensors $\underline{A}_n$ evaluated at the time and place at which the stress is to be evaluated. Moreover, all of these tensors can be evaluated directly when the velocity is given in the usual Eulerian description, as a function $\underline{u}(\underline{x},t)$ of place and time.

Let $\underline{p}(s,t,\underline{x})$ be the place at time s of the particle that is at $\underline{x}$ at time t. Then the cartesian components of the relative strain are

$$G_{ij}(s,t,\underline{x}) = p_{k,i}p_{k,j}(s,t,\underline{x}).$$

Let $\underline{u}^{(n)} = (d/ds)^n\underline{p}$. Then

$$(d/ds)^n G_{ij} = p_{k,i}u^{(n)}_{k,j} + p_{k,j}u^{(n)}_{k,i} + \sum_{m=1}^{n-1}\binom{n}{m}u^{(m)}_{k,i}u^{(n-m)}_{k,j}.$$

This is not the value of the n-th kinematic tensor $A^{(n)}_{ij}$ at time s. It is a function of s,t, and $\underline{x}$. $\underline{u}^{(1)}(s,t,\underline{x})$ is the velocity at the place $\underline{p}(s,t,\underline{x})$. $u^{(1)}_{i,j}$ is not the velocity gradient there, but the gradient with respect to the particle labels x_i.

To obtain expressions for the components $A^{(n)}_{ij}$ of $\underline{A}_n$ in terms of derivatives of the velocity $\underline{u}(\underline{x},t)$ (given as a function of place and time), we evaluate $(d/ds)^n G_{ij}$ at time $s = t$. At that time, $\underline{p} = \underline{x}$ and $p_{i,j} = \delta_{ij}$. Moreover,

$$\underline{u}^{(n)}(t,t,\underline{x}) = (D/Dt)^{n-1}\underline{u}(\underline{x},t).$$

We retain the same notation $\underline{u}^{(n)}$ for the function on the right, at the risk of some confusion. Then we obtain

$$A_{ij}^{(n)} = (u_i^{(n)})_{,j} + (u_j^{(n)})_{,i} + \sum_{m=1}^{n-1} \binom{n}{m} u_{k,i}^{(m)} u_{k,j}^{(n-m)}.$$

The spatial differentiation is carried out after the material time differentiation. For functions of s, t, and $\underline{x}$, differentiation with respect to s and $\underline{x}$ is commutative, but D/Dt does not commute with the gradient operation on $\underline{u}(\underline{x}, t)$.

Problem: Write out the terms of $(\partial/\partial x_j)(D/Dt)u_i(\underline{x},t)$ and $(D/Dt)(\partial/\partial x_j)u_i(\underline{x},t)$.

Sometimes the notation $\underline{A}_0 = \underline{I}$ is used. The components of $\underline{A}_1$ and $\underline{A}_2$ are

$$A_{ij}^{(1)} = u_{i,j} + u_{j,i}$$

and

$$A_{ij}^{(2)} = a_{i,j} + a_{j,i} + 2u_{k,i}u_{k,j},$$

where $\underline{a}$ is the acceleration, $\underline{a} = D\underline{u}/Dt$. Notice that the strain-acceleration $\underline{A}_2$ is not expressed in terms of the acceleration $\underline{a}$ in the same way that the strain-rate $\underline{A}_1$ is expressed in terms of the velocity $\underline{u}$.

It is often convenient to obtain the tensors $\underline{A}_n$ by using the following recursion relation, rather than by using the explicit expression given above:

$$A_{ij}^{(n+1)} = DA_{ij}^{(n)}/Dt + A_{ik}^{(n)}u_{k,j} + A_{jk}^{(n)}u_{k,i}.$$

The operation by which $\underline{A}_{n+1}$ is obtained from $\underline{A}_n$ is Oldroyd differentiation. The reason for the extra terms is something like the reason for the extra terms in a covariant derivative of a general tensor.

Problems:

1. Derive this recursion relation.

2. Express $\underline{A}_2$ explicitly in terms of velocity and its derivatives with respect to $\underline{x}$ and t.

3. Evaluate all tensors $\underline{A}_n$, for a steady simple shearing motion. Why is the strain-acceleration not zero although the acceleration is zero? Explain in physical or geometrical terms.

4. Evaluate the stress in steady simple shearing, correct to third order.

5. Evaluate the position $\underline{p}(s,t,\underline{x})$ for a steady flow with velocity field $\underline{u} = ay\underline{i} + bz\underline{j}$. (Use $D\underline{p}/Ds = \underline{u}$.) Evaluate the relative deformation gradient, the relative strain, and all of the Rivlin-Ericksen tensors. Evaluate the stress, correct to third order. What parameters need to be small in order to use slow-motion approximations?

7. Solution of Problems.

In order to get correct answers when using slow-motion approximations, it is necessary to use ordinary perturbation methods, starting from the Newtonian (Navier-Stokes) solution as the lowest-order approximation. In other words, it is necessary that the solution can be expanded in powers of a small parameter, with the Newtonian flow solution as the first term, and that the higher-order terms do indeed represent only a small perturbation on this first term. If a solution is obtained directly without using a perturbation procedure, it is valid only if it is possible to expand it in this way. It is necessary to mention this because the higher-order approximate constitutive equations contain higher derivatives, and they can give rise to formal "solutions" which are not small perturbations on Newtonian flow. A "solution" which does not satisfy the criterion of slowness used in formulating the constitutive equation is merely a curiosity with no meaning.

Example: Consider fluid in a channel with walls at $y = 0, H$, flowing in the x-direction with speed $u(y,t)$. The shearing stress component $\sigma = \sigma_{xy}$ is given to "second order" by

$$\sigma(y,t) = \eta_o u_y(y,t) - \eta_o T u_{yt}(y,t) + \dots .$$

(Problem: Verify this.)

Beware. We have not defined what is meant by "second order". In the absence of a pressure gradient, the x-component of the momentum equation is

$$\rho u_t = \eta_o u_{yy} - \eta_o T u_{yyt} + \dots .$$

To solve initial-value problems, we use separation of variables.

Beware. The constitutive equation is bound to be meaningless in initial-value problems, until enough time has elapsed for the initial break to have been forgotten. Proceeding formally, we find that separated solutions satisfying $u = 0$ at the walls have the form $\sin(n\pi y/H)\exp(\lambda_n t)$, where λ_n must satisfy

$$\rho\lambda_n = -\eta_o(1-\lambda_n T + \dots)(n\pi/H)^2.$$

We can now see what should be meant by "second order". The relevant dimensionless parameter that has appeared is $\lambda_n T$. In order for the constitutive equations to have any validity, it is necessary (but not sufficient) that $|\lambda_n T| \ll 1$. If we ignore this important stipulation and merely solve for λ_n as if no higher-order terms were relevant, we obtain

$$\lambda_n T = -\frac{kn^2}{1-kn^2}, \quad k = \eta_o T\pi^2/\rho H^2.$$

If we consider an initial-value problem with $u(y,0) = \epsilon \sin(n\pi y/H)$, the solution is $u(y,t) = \epsilon \sin(n\pi y/H)\exp(\lambda_n t)$. If λ_n is negative, the initial disturbance dies out and the fluid becomes quiet again, as of course it must. However, our "solution" shows that if the prescribed value of n^2 exceeds $1/k$, λ_n is positive. In that case, no matter how small the initial velocity amplitude ϵ may be, the velocity becomes arbitrarily large as time progresses, according to the formal solution. In the usual way of analyzing stability, we would conclude that the state of rest is unstable. This is all nonsense, of course. If kn^2 is greater than unity, then so is $\lambda_n T$, a parameter which must be small if the analysis is to have any meaning. If kn^2 is not small, we should use the linear viscoelastic constitutive equation. If kn^2 is small, the present analysis is all right, but the expression for λ_n should be written as

$$\lambda_n T = -kn^2 - (kn^2)^2 - \ldots .$$

The constitutive equation that we have used does not tell us what the terms of higher order in this series may be. The solution is correct to the order to approximation indicated, and it shows that within the range of validity of the solution, λ_n is negative and an initial small disturbance must die out.

Problem: Solve this problem by using the linear viscoelastic constitutive equation.

8. Ordinary Perturbations.

To avoid spurious solutions, we use ordinary perturbation procedures. Consider slow, smooth, steady motions, for which the stress is expanded in powers of a small parameter TU/L, in effect if not in fact. We expand all variables in powers of this parameter and obtain equations for the terms of the series.

Let us consider creeping flow (zero Reynolds number) so that we do not need to bother with the inertial term in the momentum equation. Then the equations of motion are

$$\underline{\nabla}\cdot\underline{u} = 0 \quad \text{and} \quad \underline{\nabla} p = \text{Div}\, \underline{S}.$$

We write the velocity and pressure as

$$\underline{u} = \underline{u}_1 + \underline{u}_2 + \ldots \quad \text{and} \quad p = p_1 + p_2 + \ldots ,$$

where the n-th term of either series is understood to be of order $(TU/L)^n$. By substituting the velocity series in the condition of incompressibility and separating terms of different orders, we find that $\underline{\nabla} \cdot \underline{u}_n = 0$. As for the momentum equation, we must first use the series expansion of the velocity field to evaluate the terms in the series for $\underline{S}$, $\underline{S} = \underline{S}_1 + \underline{S}_2 + \dots$, where $\underline{S}_n$ has the form previously discussed. Now

$$\underline{S}_1 = \underline{S}_1(\underline{u}_1) + \underline{S}_1(\underline{u}_2) + \underline{S}_1(\underline{u}_3) + \dots ,$$

where $\underline{S}_1(\underline{u}_n)$ is the contribution that the term $\underline{u}_n$ makes to the Newtonian stress $\underline{S}_1 = \eta_o \underline{A}_1$. $\underline{S}_1(\underline{u}_n)$ is of n-th order (and thus of first order, i.e. bounded in comparison to TU/L; there is no contradiction). Since $\underline{S}_2$ is non-linear in the velocity, it involves a term $\underline{S}_2(\underline{u}_1)$ that is of second order, a term $\underline{S}_2(\underline{u}_1, \underline{u}_2)$ of third order, and so on. In general, $\underline{S}_n$ can be expanded as a sum of terms of order n and higher.

Problem: Evaluate $\underline{S}_1(\underline{u}_n)$, $\underline{S}_2(\underline{u}_1)$, $\underline{S}_2(\underline{u}_1, \underline{u}_2)$, and $\underline{S}_3(\underline{u}_1)$. What is the order of $\underline{S}_2(\underline{u}_2)$, the terms of $\underline{S}_2$ that involve only $\underline{u}_2$?

When terms of various orders are separated in the momentum equation, we obtain

$$\underline{\nabla} p_1 - \eta_o \nabla^2 \underline{u}_1 = \underline{0},$$

$$\underline{\nabla} p_2 - \eta_o \nabla^2 \underline{u}_2 = \mathrm{Div}\, \underline{S}_2(\underline{u}_1),$$

$$\underline{\nabla} p_3 - \eta_o \nabla^2 \underline{u}_3 = \mathrm{Div}\, \underline{S}_2(\underline{u}_1, \underline{u}_2) + \mathrm{Div}\, \underline{S}_3(\underline{u}_1),$$

and so on. The first equation is the usual equation for Stokes flow. When $\underline{u}_1$ has been found, we can evaluate the right-hand side of the second equation, and then determine $\underline{u}_2$ and p_2, and so on. However, we must make sure that $\underline{S}_2(\underline{u}_1)$ is really only a small perturbation on $\underline{S}_1(\underline{u}_1)$. If not, we are using the wrong constitutive equation, and the whole procedure is meaningless.

9. A Useful Identity.

The right-hand side of the second-order equation can be simplified a little. To simplify the notation, let us write $\underline{v} = \underline{u}_1$ for the Newtonian velocity field, $\underline{A} = \underline{A}_1(\underline{u}_1)$, and $\underline{B} = \underline{A}_2(\underline{u}_1)$. Then,

$$\underline{S}_2(\underline{u}_1) = -\eta_o T(\underline{B}-\underline{A}^2) + \eta_o T^*\underline{A}^2.$$

Let us consider the divergence of the term $\underline{B}-\underline{A}^2$. We can prove that if $\text{Div}\,\underline{A}$ (i.e. $\nabla^2\underline{v}$) is irrotational and $\underline{\nabla}\cdot\underline{v} = 0$, then $\text{Div}(\underline{B}-\underline{A}^2)$ is also irrotational. If P is the potential for $\text{Div}\,\underline{A}$, so that $\text{Div}\,\underline{A} = \underline{\nabla}P$, then

$$\text{Div}(\underline{B}-\underline{A}^2) = \underline{\nabla}\left(\frac{DP}{Dt} + \frac{1}{2}\gamma^2\right).$$

Here γ is the absolute shear rate, defined by

$$\gamma^2 = \frac{1}{2}\text{tr}(\underline{A}^2) = \frac{1}{2}\text{tr}\underline{B}.$$

(Problem: Prove that $\text{tr}\underline{B} = \text{tr}(\underline{A}^2)$ if $\underline{\nabla}\cdot\underline{v} = 0$.) Of course, in creeping flow the potential P is just p_1/η_o.

The theorem is proved by rearranging the expression for $\text{Div}\,\underline{B}$. Now, $\underline{A}$ and $\underline{B}$ are given in terms of $\underline{v}$ by

$$A_{ij} = v_{i,j} + v_{j,i}$$

and

$$B_{ij} = DA_{ij}/Dt + A_{ik}v_{k,j} + A_{jk}v_{k,i},$$

where, of course, $D/Dt = \partial/\partial t + v_k\partial/\partial x_k$. Hence we obtain

$$B_{ij,j} = D(A_{ij,j})/Dt + v_{k,j}A_{ij,k} + (A_{ik}v_{k,j})_{,j}$$

$$+ A_{jk,j}v_{k,i} + A_{jk}v_{k,ij}.$$

Problems:

1. Show that if $v_{i,i} = 0$ as assumed, the second and third terms combine to yield $\text{Div}\ \underline{A}^2$.

2. Show that the final term is equal to $\frac{1}{2}\underline{\nabla}\,\gamma^2$.

3. Show that if $\text{Div}\ \underline{A} = \underline{\nabla}P$, as assumed, then the first and fourth terms combine to give $\underline{\nabla}(DP/Dt)$.

4. Rearrange the equation so as to give the desired expression for $\text{Div}(\underline{B}-\underline{A}^2)$.

In the case of creeping flow, for which $P = p_1/\eta_o$, the second-order momentum equation now takes the form

$$\underline{\nabla}(p_2 + T\ Dp_1/Dt + \tfrac{1}{2}\eta_o T\gamma^2) - \eta_o\nabla^2\underline{u}_2 = \eta_o T^*\text{Div}\ \underline{A}^2.$$

From this equation we can see that p_2 will be of the form

$$p_2 = -T\ Dp_1/Dt - \tfrac{1}{2}\,\eta_o T\gamma^2 + p_2^*,$$

where p_2^* is whatever is needed to balance the term involving $\underline{u}_2$ and the right-hand side. That part of p_2 involving the coefficient T is now explicitly known.

The stress is now given to second order by

$$\underline{\sigma} = -\underline{I}[(p_1 - T\ Dp_1/Dt) - \tfrac{1}{2}\eta_o T\gamma^2 + p_2^*] + \eta_o[\underline{A} - T(\underline{B}-\underline{A}^2)].$$

Some of this has a simple interpretation. Because of memory effects, stress is convected. The stress is mainly just what we would obtain for Newtonian flow, but a fluid element retains some of the stress that it had in the past as it moves along. The stress in a material element at time t is in part something like the Newtonian stress at that particle at time $t-T$. The combination $p_1 - T\,Dp_1/Dt$ is the value of p_1 at the place where the particle was at time $t-T$, to the order of approximation considered here. Similarly, $\underline{A}-T\underline{B}$ has something to do with the value of $\underline{A}$ at the earlier place. Using $\underline{B}$ instead of $D\underline{A}/Dt$ tends to take into account the rotation of the built-in stress field as the particle rotates. Since $\mathrm{tr}\underline{A} = 0$, a rotated version of $\underline{A}$ would still be deviatoric. $\underline{B}$ is not deviatoric itself, but $\underline{B}-\underline{A}^2$ is. That is roughly why $\underline{B}-\underline{A}^2$ appears in place of a simple time-derivative of $\underline{A}$.

Thus, the stress field is in part a distorted version of the Newtonian stress field. When the Newtonian stress has been warped by convection, it is no longer quite in equilibrium. The extra term $-\frac{1}{2}\eta_0 T\gamma^2$ in the pressure is apparently all that is needed to equilibrate the Newtonian field after it has been distorted.

10. Plane Flow.

In plane flow with $\underline{\nabla}\cdot\underline{v} = 0$, $\underline{A}^2$ is essentially a 2×2 matrix, and it has the form

$$\underline{A}^2 = \gamma^2(\underline{I} - \underline{kk}).$$

(Problem: Verify.)

Then $\mathrm{Div}\,\underline{A}^2 = \underline{\nabla}\gamma^2$. Consequently, the coefficient of T^* in the second-order momentum equation is equivalent to a pressure gradient.

It follows that the second-order momentum equation can be satisfied with $\underline{u}_2 = \underline{0}$. In other words, the Newtonian velocity field is still correct to second order. The term in p_2 that we called p_2^* is equal to $\eta_0 T^*\gamma^2$, and the total stress is that previously discussed, plus $-\underline{I}\eta_0 T^*\gamma^2$. Consequently, to solve plane creeping flow problems with velocity boundary conditions correct to second order, we need to find only the Newtonian velocity field. (Tanner's Theorem).

As an application, let us consider a steady simple shearing flow over a plane wall $y = 0$, disturbed by a deep, narrow slot in the wall. Let $x = 0$ be the slot centerline, and let $x = \pm L$, for y negative, be the walls of the slot. Far from the slot mouth for y positive, the velocity is asymptotic to $\gamma_0 y \underline{i}$, the velocity that would obtain everywhere if the flow were not disturbed. The velocity deep inside the slot approaches zero.

The reason for considering this problem is that normal stresses are often measured by attaching a stand-pipe to a small hole in a wall that the fluid is flowing over. The guage pressure p_g is assumed to be equal to the thrust that the fluid would exert on the wall if the hole were not there, p_u.

If the Reynolds number defined by $\rho(\gamma_0 L)L/\eta_0$ is sufficiently small, the creeping flow approximation is applicable. Now, Newtonian creeping flow is reversible, so the streamlines are symmetrical about the slot axis $x = 0$. If u and v are the velocity components, then u is an even function of x and v is odd. It follows that v, v_{xx}, and v_{yy} all vanish on $x = 0$, so $\nabla^2 v = 0$ there. From the y-component of the momentum equation, $\partial p_1/\partial y = \eta_0 \nabla^2 v$, we conclude that p_1 is constant along the line $x = 0$. Then $p_u = p_1(0,\infty)$ and $p_g = p_1(0,-\infty)$ are the same. Thus, if the hole is small enough that the zero-Reynolds-number approximation is valid, the undisturbed pressure and the guage pressure are the same, in the Newtonian approximation.

Now consider the second-order approximation. Far from the slot, where the motion is undisturbed and $p_1 = p_\infty$, say, the stress component σ_{yy} is

$$-p_u = \sigma_{yy} = -p_\infty - \frac{1}{2}\eta_0 T \gamma_0^2.$$

This is what the pressure of the fluid against the wall would be if there were no disturbance. Deep in the slot, there is no motion and $p_1 = p_\infty$ still, so the guage pressure is $p_g = p_\infty$. Hence, in the second order of approximation, the guage pressure is not equal to the undisturbed thrust of the fluid against the wall:

$$p_g - p_u = -\frac{1}{2}\eta_o T\gamma_o^2.$$

The guage reads low, and once the hole is small enough to avoid inertial effects, the error is independent of hole size.

The error occurs because in a shearing motion, viscoelastic effects produce an extra tension along the direction of relative sliding motion, which is often roughly in the direction of the streamlines. In Newtonian flow, the streamlines dip in a little as they cross the mouth of the slot. By looking at the equilibrium of a control volume with pressure p_u on one end and p_g on the other, we see that the extra tension partly balances p_u.

11. Flow in Tubes.

We found that in plane creeping flow, the Newtonian velocity field still satisfies the momentum equation to second order. The normal stresses change in second order, but the velocity field does not. This is roughly because normal stresses are even functions of the shear rate while shearing stresses are odd. In third order the apparent viscosity is not the same as in Newtonian flow, so the Newtonian velocity field is no longer correct to third order.

Steady flow in tubes is another class of problems in which we can see this kind of effect. Consider a tube with axis along the z-direction, with cross-sectional boundary C (not necessarily circular). In the Newtonian approximation we can have steady flow in straight lines, $\underline{v} = w(x,y)\underline{k}$. The stress σ_{zz} is constant over each cross-section, and varies linearly in the z-direction, $\partial\sigma_{zz}/\partial z = G$. G is the magnitude of the pressure gradient. The z-component of the momentum equation yields

$$\nabla^2 w = -G/\eta_o.$$

The boundary condition is $w = 0$ on C.

Dimensional inspection shows that $U = GL^2/\eta_o$ is a characteristic velocity; here L is a typical length associated with the cross-sectional shape, such as the radius if the cross-section is circular. A typical shear rate is $U/L = GL/\eta_0$, so the amount of shear in one relaxation time is of the order of $A = TU/L = TGL/\eta_0$. Thus we can use the slow-motion approximations if the pressure gradient or the tube diameter is sufficiently small.

The Newtonian pressure p_1 is $-Gz$, plus a constant. Since $\nabla^2\underline{v} = \underline{\nabla}P$, with $P = -Gz/\eta_0$ in this case, $\mathrm{Div}(\underline{B}-\underline{A}^2)$ is irrotational. Consider $\mathrm{Div}\,\underline{A}^2$. We find that

$$\underline{A} = \underline{\nabla}\,\underline{v} + (\underline{\nabla}\,\underline{v})^t = \underline{\nabla}w\,\underline{k} + \underline{k}\,\underline{\nabla}w,$$

so

$$\underline{A}^2 = \underline{\nabla}w\,\underline{\nabla}w + \underline{k}\underline{k}\gamma^2, \quad \gamma^2 = \underline{\nabla}w \cdot \underline{\nabla}w.$$

Here we have used $\underline{k}\cdot\underline{\nabla}w = 0$. Then,

$$\begin{aligned}\mathrm{Div}\,\underline{A}^2 &= (\underline{\nabla}\cdot\underline{\nabla}\,w)\underline{\nabla}w + (\underline{\nabla}w\cdot\underline{\nabla})\underline{\nabla}w + (\underline{\nabla}\cdot\underline{k})\underline{k}\gamma^2 + (\underline{k}\cdot\underline{\nabla})\underline{k}\gamma^2 \\ &= -(G/\eta_0)\underline{\nabla}w + \tfrac{1}{2}\underline{\nabla}(\gamma^2).\end{aligned}$$

(In index notation, $(\underline{\nabla}w\cdot\underline{\nabla})\underline{\nabla}w$ gives $w_{,j}w_{,ij} = \frac{1}{2}(w_{,j}w_{,j})_{,i}$.) Thus, $\mathrm{Div}\,\underline{A}^2$ is the gradient of a scalar.

It follows that the second-order momentum equation is still satisfied with $\underline{u}_2 = \underline{0}$, i.e. the Newtonian velocity field is still all right to second order. The term p_2^* in the pressure perturbation is

$$p_2^* = \eta_0 T^*[-(Gw/\eta_0) + \tfrac{1}{2}\gamma^2].$$

At a point on the boundary C where the unit normal is $\underline{n}$, the thrust of fluid against the wall is given to second order by

$$-\underline{n}\cdot\sigma\underline{n} = p_1 + p_2 - \eta_0\underline{n}\cdot[\underline{A}-T(\underline{B}-\underline{A}^2)]\underline{n} - \eta_0 T^*\underline{n}\cdot\underline{A}^2\underline{n}.$$

This reduces to

$$-\underline{n}\cdot\sigma\underline{n} = -Gz + \frac{1}{2}(T-T^*)\eta_0\gamma^2 + \text{const.}$$

Problems:

1. Verify this.

2. Find the Newtonian flow in a tube of equilateral triangular cross-section. (The speed $w(x,y)$ is a polynomial of low degree in this case.) Evaluate the wall thrust and draw a graph showing how it varies along the boundary.

To measure the wall thrust one might attach a pressure guage to a slot in the wall of the tube. If the slot mouth is parallel to the tube axis and very long, the tube plus slot is just a new tube of a different cross-section. The guage pressure p_g is the value of $-\underline{n}\cdot\sigma\underline{n}$ deep inside the slot, where the fluid is nearly motionless if the slot is thin. The undisturbed pressure p_u which would be present if the slot were absent, which is what one wants to measure, is not the same; we get

$$p_g - p_u = -\frac{1}{2}(T-T^*)\eta_0\gamma^2,$$

where γ is the shear rate at the site of the slot mouth in the undisturbed flow.

Normal stress differences in shearing flow, given as functions of the shear rate, are basic response functions for viscoelastic fluids. A considerable amount of effort has gone into measuring them. In the present problem and the preceding one (plane flows) we have found that undisturbed wall thrust and guage pressure are not the same thing. Because these pressure-hole errors were only recently discovered, much of the literature giving data on normal stress differences is incorrect. The data itself is presumably correct, but its interpretation in terms of normal stresses is wrong.

Let's now consider the third-order momentum equation for flow in tubes. Since $\underline{u}_2 = \underline{0}$, the term $\text{Div } \underline{S}_2(\underline{u}_1, \underline{u}_2)$ goes out, and we are left with

$$\nabla p_3 - \eta_0 \nabla^2 \underline{u}_3 = \text{Div } \underline{S}_3(\underline{u}_1).$$

$\underline{S}_3(\underline{u}_1)$ is a perturbation on the Newtonian shearing stress $\eta_0 \underline{A}$ due to the fact that the apparent viscosity is $\eta(\gamma^2)$ rather than $\eta(0)$.

(Problem: Show that with the form of $\underline{S}_3$ given earlier, $\underline{S}_3$ is equal to $2(C_2+C_3)\gamma^2\underline{A}$ when $\underline{u}_1 = \underline{k}w(x,y)$. Conclude that $d\eta(\gamma^2)/d(\gamma^2)$, evaluated at $\gamma = 0$, is equal to $2(C_2+C_3)$.)

The right-hand side of the equation is a vector in the direction $\underline{k}$, so the equation can be satisfied with $\underline{u}_3 = \underline{k}w_3(x,y)$ and $p_3 = 0$. Thus in this order, the axial velocity is perturbed but normal stresses are not.

(Problem: Set up the equation for w_3. The integral of w_3 over the cross-section is the efflux perturbation. Use the equation for w_3 and the divergence theorem to express the efflux perturbation directly in terms of the Newtonian shear rate.)

In fourth order we should expect a perturbation on the normal stresses but not the velocity. Actually, the axial velocity is not changed in fourth order, but normal stress differences cause the fluid to circulate around the cross-section, and the flow is no longer in straight lines. Since the transverse circulation is very weak (a fourth order effect), it is hard to observe, but Kearsley has measured it by photographing dye streaks. Langlois and Rivlin have done the calculation (Rendiconti di Matematica $\underline{22}$, 169 (1963)).

CHAPTER IX

VISCOMETRIC FLOW

On the edge of the flow diagnosis diagram where $\omega T = 0$, the shear amplitude A is γT, the amount of shear in one relaxation time. We can inch away from the Newtonian corner by using slow viscoelastic flow approximations, but as we have seen in the case of tube flow, this can become very laborious. We need a constitutive equation that deals directly with steady shearing motions at arbitrary shear rates.

If a material element is doing nothing but shearing at a constant rate (plus translating and rotating) it is a simple matter to say how the stress in it can depend on the motion. Moreover, there are a few special classes of flows, called viscometric flows, in which all material elements are shearing steadily. In the present chapter we first describe some kinematically admissible examples of viscometric flow and briefly mention some cases in which exact solutions can be obtained. In the remainder of the chapter we discuss some motions that are not exactly viscometric, mainly in qualitative terms. The object is to show how to make qualitative predictions about steady laminar shearing motions and to indicate methods of approximation that might be used in obtaining rough quantitative estimates.

1. Stress.

Most of the flows along the edge $\omega T = 0$ are viscometric flows, in which each material element is undergoing simple shearing motion (plus translation and rotation) with a constant rate of shear. Different material elements can be shearing at different rates, and indeed it may happen that some material elements in a given flow are in viscometric motion while others are not. However, let us begin with the simplest kind of viscometric flow, steady simple shearing motion itself, with velocity $\underline{u} = \gamma y \underline{i}$.

The stress components are functions of the shear rate γ. We can say a little about these functions without doing any experiments, if the material is isotropic. Someone standing on the other side of the experimental apparatus and using his own coordinate system would say that the shear rate is $-\gamma$. He would see the same stresses, but assign different signs to some of them. Since his report of the situation must agree with ours, we can deduce that σ_{xy} is an odd function of the shear rate, σ_{xz} and σ_{yz} are zero, and the normal stress components are even functions of the shear rate. Assuming that the fluid is incompressible, the normal stresses are indeterminate to the extent of a common pressure, so only the normal stress differences are definite functions of the shear rate. We will use the notation

$$\sigma_{xy} = \gamma\eta(\gamma^2), \quad \sigma_{xx}-\sigma_{yy} = \gamma^2 N_1(\gamma^2), \quad \sigma_{yy}-\sigma_{zz} = -\gamma^2 N_2(\gamma^2).$$

If $\gamma = 0$ there has never been any motion, since we are talking about cases in which γ is constant in time. In that case the stress must reduce to an isotropic pressure, as the notation indicates. $\sigma_{xx}-\sigma_{yy}$ is usually called the <u>first</u> normal stress difference, and $\sigma_{yy}-\sigma_{zz}$ is the <u>second</u> normal stress difference. The normal stress coefficient N_1 is positive for polymer solutions, and N_2 is very likely positive also (so that the second normal stress difference is <u>negative</u>). N_2 is small in comparison to N_1, 10 percent or so. Thus, normal stress effects in steady simple shearing amount to an extra tension in the direction of shearing, plus a relatively small extra pressure on planes y = constant.

Instead of writing $\underline{i}$, $\underline{j}$, and $\underline{k}$ for the base vectors here, let us write $\underline{a}$, $\underline{b}$, and $\underline{c}$. The direction of relative sliding motion is $\underline{a}$, the normal to the slip surfaces is $\underline{b}$, and $\underline{c}$ is orthogonal to the other two. We will use the same notation in more complicated examples in which $\underline{a}$, $\underline{b}$, and $\underline{c}$ vary with position. In terms of these vectors, the stress is

$$\underline{\sigma} = -p\underline{I} + \gamma\eta(\underline{ab} + \underline{ba}) + \gamma^2(N_1-N_2)\underline{aa} - \gamma^2 N_2\underline{bb}.$$

2. Example: Channel Flow.

Consider flow in the x-direction in a channel with walls at $y = \pm L$, the velocity being $\underline{u} = u(y)\underline{i}$. The material on each plane y = constant moves all together as a rigid material surface sliding tangentially. The slip surfaces just above and below it are moving at slightly different rates, so a volume element cut by the slip surface y = constant is shearing at the rate $\gamma = u'(y)$, as well as translating. With $\underline{a}$, $\underline{b}$, and $\underline{c}$ the same as the base vectors $\underline{i}$, $\underline{j}$, and $\underline{k}$, the stress is given by the expression above, in which γ is now a function of y.

Since there is no acceleration, the momentum equation reduces to

$$\underline{\nabla} p = \underline{a}\, d(\gamma\eta)/dy - \underline{b}\, d(\gamma^2 N_2)/dy.$$

It follows that

$$p = -Gx - \gamma^2 N_2 + \text{const.}$$

and

$$d(\gamma\eta)/dy = -G.$$

Here the constant of integration G is the magnitude of the axial component of the pressure gradient.

We use the no-slip condition $u(\pm L) = 0$, so we anticipate that $u(y)$ is an even function of y, and γ is odd, so

$$\gamma\eta = -Gy.$$

Let $\gamma = \Gamma(\sigma)$ be the inverse of $\sigma = \gamma\eta(\gamma^2)$. Then

$$u'(y) = \gamma = -\Gamma(Gy).$$

The total flux Q is of more interest than the detailed velocity profile:

$$Q = \int_{-L}^{L} u\,dy = yu\Big|_{-L}^{L} - \int_{-L}^{L} yu'\,dy$$

$$= \int_{-L}^{L} y\Gamma(Gy)dy = (2/G^2)\int_{0}^{GL} \sigma\Gamma(\sigma)d\sigma.$$

By plotting experimental values of G^2Q versus GL, we would get a curve with slope $2GL\Gamma(GL)$, from which values of Γ and thus eventually η could be deduced. Of course, Poiseuille flow (pipe flow) would be more convenient than channel flow for experiments. The analogous relation for pipe flow is called the Weissenberg-Rabinowitsch-Mooney formula, because Mooney said that Weissenberg said that Rabinowitsch said how to do it. Mooney also showed how to account for wall slip.

Problem: Derive the relation between flux and pressure gradient for a pipe of radius R.

3. Slip Surfaces, Shear Axes, and Shear Rate.

Let us consider some examples of viscometric flow. We will discuss only the kinematics, and not worry yet about whether or not these flows satisfy the momentum equations.

All viscometric flows can be visualized as the relative sliding motion of inextensible material surfaces, which we call slip surfaces. The direction of relative sliding motion of two neighboring slip surfaces is denoted $\underline{a}$; $\underline{a}$ is a unit vector. The direction normal to a slip surface is $\underline{b}$, another unit vector. Let $\underline{c}$ be a unit vector normal to $\underline{a}$ and $\underline{b}$. We call $\underline{a}$, $\underline{b}$, and $\underline{c}$ the shear axes. They can vary in direction from one particle to another, and also change in time at a single particle.

In steady parallel flows, $\underline{u} = u(x,y)\underline{k}$, the slip surfaces are the general cylinders $u = \text{const}$. Each slip surface moves as if rigid, sliding tangentially in the axial direction at a constant speed. The direction of relative sliding motion is $\underline{a} = \underline{k}$ (or $-\underline{k}$, it doesn't matter which we use). The normal direction $\underline{b}$ is parallel to $\underline{\nabla}u$. The magnitude of $\underline{\nabla}u$ is the shear rate γ. At any given particle the shear rate is constant in time, which is essential. We note that the velocity gradient is $\underline{\nabla}\,\underline{u} = \gamma\underline{b}\,\underline{a}$.

Another kind of velocity field in which all particles move in straight lines is

$$\underline{u} = u(z)\underline{i} + v(z)\underline{j}.$$

The slip surfaces are the parallel planes $z = \text{const}$. Each moves as if rigid. The normal direction is $\underline{b} = \underline{k}$. The shear rate and direction of relative sliding $\underline{a}$ are given together by

$$\gamma\underline{a} = \underline{u}'(z) = u'(z)\underline{i} + v'(z)\underline{j}.$$

In this case the $\underline{a}$-direction is not the direction of motion. In fact, if the speed is independent of z, $\underline{a}$ is perpendicular to $\underline{u}$ everywhere. Notice that $\underline{\nabla}\,\underline{u} = \underline{\nabla} z\, \underline{u}'(z) = \gamma\underline{b}\,\underline{a}$.

Steady circular motions with velocity $\underline{u} = r\omega(r,z)\underline{i}_\theta$ are also viscometric. The slip surfaces are the material surfaces of revolution on which the angular velocity ω is constant. The direction $\underline{a}$ is $\underline{i}_\theta$, the same as the direction of motion in this case. Notice that the $\underline{a}$-direction at a given particle rotates steadily. The shear axes are not constant in either time or space. The normal direction $\underline{b}$ is parallel to $\underline{\nabla}\omega$. $\underline{c}$ is tangential to the slip surface, in the meridianal direction. To find the shear rate, let us calculate $\underline{\nabla}\,\underline{u}$. We get

$$\begin{aligned}\underline{\nabla}\,\underline{u} &= \underline{\nabla}(r\omega)\underline{i}_\theta + r\omega\underline{\nabla}\theta\ \underline{i}'_\theta(\theta)\\ &= r\underline{\nabla}\,\omega\underline{i}_\theta + \omega(\underline{i}_r\underline{i}_\theta - \underline{i}_\theta\underline{i}_r).\end{aligned}$$

From the previous examples we might have expected to get something of the form $\gamma\,\underline{b}\,\underline{a}$. The first term has this form if we identify γ as $r|\underline{\nabla}\omega|$. The second term, antisymmetric, is the velocity gradient for a rotation at the angular velocity ω.

In less obvious cases we eliminate the rotation by considering $\underline{A}_1 = \underline{\nabla}\,\underline{u} + (\underline{\nabla}\,\underline{u})^t$.

If the slip surfaces are coaxial circular cylinders they can both rotate about their common axis and translate along it, giving a motion with helical streamlines:

$$\underline{u} = r\omega(r)\underline{i}_\theta + u(r)\underline{i}_z.$$

The normal direction is radial, $\underline{b} = \underline{i}_r$. The velocity gradient is

$$\underline{\nabla}\,\underline{u} = \underline{\nabla} r(r\omega'\underline{i}_\theta + u'\underline{i}_z) + \omega\underline{\nabla} r\,\underline{i}_\theta + r\omega\underline{\nabla}\theta\,\underline{i}'_\theta(\theta)$$

$$= \underline{i}_r(r\omega'\underline{i}_\theta + u'\underline{i}_z) + \omega(\underline{i}_r\underline{i}_\theta - \underline{i}_\theta\underline{i}_r).$$

The second term is evidently due to rigid rotation. Identifying the first as $\gamma\underline{b}\,\underline{a}$, we find that

$$\gamma\underline{a} = r\omega'(r)\underline{i}_\theta + u'(r)\underline{i}_z.$$

The magnitude γ is evidently a function of r only, and each particle stays on the same cylinder r at all times, so γ is constant in time at each particle, as required.

Flows with coaxial helicoidal slip surfaces have velocity fields of the form

$$\underline{u} = (r\underline{i}_\theta + c\underline{i}_z)\omega(r, z-c\theta).$$

In these flows the streamlines are helices, all with the same rise per turn, $2\pi c$. (In the preceding example the slip surfaces were cylinders, not helicoids, and the streamlines on different cylinders could have different pitches .) In these flows $\underline{a}$ is parallel to $\underline{u}$ and $\underline{b}$ is parallel to $\underline{\nabla}\omega$. The velocity gradient is

$$\underline{\nabla}\,\underline{u} = \underline{\nabla}\omega(r\underline{i}_\theta + c\underline{i}_z) + \omega(\underline{i}_r\underline{i}_\theta - \underline{i}_\theta\underline{i}_r).$$

Identifying the first term as $\gamma\underline{b}\,\underline{a}$, we find that the shear rate is given by

$$\gamma^2 = (r^2+c^2)\underline{\nabla}\,\omega\cdot\underline{\nabla}\omega,$$

and we see that it is constant in time at each particle.

There are a great many other kinds of viscometric flows, but the others appear to be impossible to produce in the laboratory. Some have slip surfaces that curl up (slip surfaces are inextensible but not necessarily rigid, as a general kinematical matter), and some are globally unsteady although the shear rate is constant at each particle. There are also more reasonable cases involving steady flow with rigid slip surfaces which nevertheless correspond to nothing that can be produced dynamically. The special flows that we have considered here are not all dynamically admissible by any means, but each class contains some flows that are.

4. Dynamical Problems.

In the preceding examples there are usually only one or two undetermined functions, depending on the coordinates in a restricted way, in the expression for the velocity field. In a given problem in which one of these flows is kinematically admissible, the undetermined functions are found by using only one or two components of the momentum equation, and the apparent viscosity function is the only material property involved in determing them. Once these functions have been determined, the remaining component(s) of the momentum equation are either automatically satisfied by some miracle (involving symmetry), or they are not. If not, we do not have a solution.

There are some special flows, such as Poiseuille and Couette flows, for which all equations are satisfied exactly. Most of these exact solutions correspond to motions in viscometers, which is why they are called viscometric. If we restrict attention to exact solutions, we will be able to use data from one kind of viscometer to tell us how some other kind of viscometer operates, but that is fairly pointless. From viscometric experimental data we wish to be able to solve problems in which the motion is not viscometric at all. Necessarily, we will find only approximate solutions, but so far as physical reality is concerned we never find anything but approximate solutions anyway.

5. Flow in Tubes.

We discussed flow in tubes earlier, using slow viscoelastic flow approximations. Let's now consider flow at an arbitrary rate. Rectilinear flows $\underline{u} = \underline{k}u(x,y)$ are kinematically admissible, and viscometric. However, remember that it was mentioned earlier that at low shear rates (low γT) the flow is in straight lines only approximately. Thus if we simply assume at the outset that the flow is rectilinear, so that we can use the viscometric expression for the stress, we will not obtain an exact solution. (Flows in tubes of circular or annular cross-section are exceptional.)

With $\underline{a} = \underline{k}$ and $\gamma\underline{b} = \underline{\nabla}u$, the stress is

$$\underline{\sigma} = -p\underline{I} + \eta(\underline{k}\underline{\nabla}u + \underline{\nabla}u\underline{k}) + \underline{\nabla}u\cdot\underline{\nabla}u(N_1 - N_2)\underline{kk} - N_2\underline{\nabla}u\,\underline{\nabla}u,$$

where η, N_1, and N_2 are given functions of $\gamma^2 = \underline{\nabla}u\cdot\underline{\nabla}u$. Since there is no acceleration, the momentum equation becomes

$$\underline{\nabla}p = \underline{\nabla}\cdot(\eta\underline{\nabla}u)\underline{k} - \underline{\nabla}\cdot(N_2\underline{\nabla}u\,\underline{\nabla}u).$$

Since the right-hand member is independent of z, then $\underline{\nabla}\partial p/\partial z = \underline{0}$, so $\partial p/\partial z$ is a constant, $-G$, say. Thus,

$$p = -Gz + P(x,y).$$

The z-component of the momentum equation is then

$$\underline{\nabla}\cdot(\eta\underline{\nabla}u) = -G,$$

and the remainder becomes

$$\underline{\nabla}P = -\underline{\nabla}\cdot(N_2\underline{\nabla}u\underline{\nabla}u)$$

$$= -\underline{\nabla}\cdot(N_2\underline{\nabla}u)\underline{\nabla}u - N_2\gamma\underline{\nabla}\gamma.$$

(Problem: Verify this.)

The last term is itself a gradient, so

$$p = -Gz - \int_0^{\gamma} N_2(\gamma^2)\gamma d\gamma + P_1(x,y),$$

where the left-over part P_1 must satisfy

$$\underline{\nabla}P_1 = -\underline{\nabla}\cdot(N_2\underline{\nabla}u)\underline{\nabla}u.$$

Thus $\underline{\nabla}P_1$ is parallel to $\underline{\nabla}u$, so surfaces P_1 = constant must coincide with surfaces u = constant. Consequently, we can write P_1 as a function of u, $P_1(u)$, and the equation then yields

$$P_1'(u) = -\underline{\nabla}\cdot(N_2\underline{\nabla}u).$$

The z-component equation gave us an equation for u, involving the apparent viscosity and the axial pressure gradient G. With $u = 0$ on the tube walls, this equation presumably determines u. Then in the equation for P_1, the right-hand side is already known. If the right-hand side turns out to be constant over surfaces u = constant, we can integrate to find P_1, but if not, we have arrived at a contradiction to the hypothesis that the flow is a parallel flow.

For a circular or annular cross-section, $u = u(r)$, so there is no contradiction. The same is true in channel flow with $u = u(y)$. In these cases an exact solution can be obtained no matter what function $N_2(\gamma^2)$ may be, as shown earlier. It is possible to show that for any other cross-section, no exact solution can be obtained except for special choices of the response function N_2.

One special choice that works is simply $N_2 = 0$. If the second normal stress difference is always zero, then P_1 is just an arbitrary constant. The statement that N_2 is zero is called the Weissenberg hypothesis. In fact N_2 is probably not zero in any case in which viscoelastic effects are of importance, but it does seem to be small in comparison to N_1.

Another special choice is $N_2 = T_2\eta(\gamma^2)$, N_2 a constant multiple of the apparent viscosity. This is qualitatively not bad. It seems that N_2 probably decreases as the shear rate increases, as η does. If we choose the proportionality constant so that $N_2 = T_2\eta$ exactly at the highest shear rate involved in the flow, the approximation should be pretty good. The reason for doing this at all is that in view of the equation that u satisfies, we will get

$$P_1'(u) = -\underline{\nabla}\cdot(T_2\eta\underline{\nabla}\, u) = GT_2,$$

so

$$P_1 = GT_2u + \text{const.}$$

Hence,

$$p = -Gz - \int_0^{\gamma} N_2(\gamma^2)\gamma d\gamma + GT_2u(x,y) + \text{const.}$$

Problems:

1. Show that at sufficiently low rates of shear this result reduces to what was previously found in the second-order slow-motion approximation. Compare stress components; p does not necessarily mean the same thing in different equations.

2. Show that the velocity field $\underline{u} = (U/\theta_0)\theta\underline{k}$ is viscometric. Deter-

mine the shear axes and the shear rate. Show that the momentum equations can be satisfied exactly, no matter what functions η, N_1, and N_2 are, by determing the reaction pressure appropriately. Evaluate the tractions on the boundaries, assuming that the fluid occupies the region $r_o \leq r \leq r_1$, $0 \leq \theta \leq \theta_o$.

6. Viscometric Equation in Terms of Rivlin-Ericksen Tensors.

A motion is viscometric only if it can be represented as the relative sliding motion of inextensible material surfaces. (This is necessary, not sufficient.) Many laminar shearing flows present this general appearance. Even if such a flow is not exactly viscometric, it may be possible to evaluate the stress with good accuracy by using the viscometric constitutive equation. However, a problem arises when we try to evaluate the shear axes and shear rate. If the motion is not exactly viscometric, these quantities are not well-defined.

We overcome this difficulty by re-writing the viscometric equation in terms of the Rivlin-Ericksen tensors. The latter can be evaluated unambiguously for any flow. The disadvantage of doing this is that in the resulting equation there will be no internal evidence that the flow should be nearly viscometric in order for the equation to be applicable, so the equation is likely to be misused.

In flows that are exactly viscometric, all Rivlin-Ericksen tensors except $\underline{A}_1$ and $\underline{A}_2$ are zero, and these two are given in terms of the shear axes and shear rate by

$$\underline{A}_1 = \gamma(\underline{a}\,\underline{b} + \underline{b}\,\underline{a}) \quad \text{and} \quad \underline{A}_2 = 2\gamma^2\underline{b}\,\underline{b}.$$

It follows that

$$\underline{A}_1^2 = \gamma^2(\underline{a}\,\underline{a} + \underline{b}\,\underline{b}) \quad \text{and} \quad \gamma^2 = \tfrac{1}{2}\mathrm{tr}\,\underline{A}_1^2.$$

Thus,

$$\gamma^2\underline{b}\,\underline{b} = \tfrac{1}{2}\,\underline{A}_2, \quad \gamma^2\underline{a}\,\underline{a} = \underline{A}_1^2 - \tfrac{1}{2}\underline{A}_2, \quad \gamma(\underline{a}\,\underline{b} + \underline{b}\,\underline{a}) = \underline{A}_1,$$

and so we can write the expression for the stress as

$$\underline{\sigma} = -p\underline{I} + \eta\underline{A}_1 - \frac{1}{2}N_1\underline{A}_2 + (N_1-N_2)\underline{A}_1^2.$$

This way of expressing the stress in terms of $\underline{A}_1$ and $\underline{A}_2$ is by no means unique. For example, we know that $\text{tr}\underline{A}_2 = \text{tr}\underline{A}_1^2$ in any isochoric flow, so either may be used to evaluate γ^2. More importantly, there are also identities involving $\underline{A}_1$ and $\underline{A}_2$ which are satisfied if the motion is viscometric but not otherwise. Consequently, there are many ways of writing the stress in terms of $\underline{A}_1$ and $\underline{A}_2$ that all give the same answer if the flow is viscometric, but each one gives a different extrapolation to non-viscometric flows. There is no point in worrying about this if the flow really is nearly viscometric, but there is plenty of reason to worry if it is not. The history is rheology is plagued with constitutive equations applied blindly to all kinds of flows in which they could have no relevance.

Notice that the equation now looks like the second-order approximation for slow viscoelastic flow. Of course, the coefficients are functions of the shear rate here, while in the second-order equation they are constants. This similarity has an interesting effect in turning wrong answers into right answers sometimes. Some people like to solve equations without worrying about the relevance of the equations to physical situations, so a great many problems have been solved on the basis of the second-order equations. Sometimes the answers seem to be right even at shear rates so high that the second-order approximation must definitely be wrong. Apparently, in steady laminar shearing motions the viscometric equation gives roughly the right stress, even if the coefficients are treated as constants. Of course, the constants used should be values of the viscometric functions at a shear rate appropriate to the problem considered. If one solves the problem first and then later chooses the values of the coefficients that give best agreement with experiment, it makes it look like the second-order approximation is working, far outside its range of validity.

Problems:

1. Express η_o, T, and T* (in the second-order slow-motion approximation) in terms of the limiting values of the viscometric functions at zero shear rate.

2. We may assume that the velocity field for flow in a long tube of non-circular cross-section has the form

$$\underline{u} = u(x,y)\underline{i} + v(x,y)\underline{j} + U(x,y)\underline{k},$$

where u and v are small in comparison to U. The main axial flow $U\underline{k}$ is, by itself, viscometric, but the flow disturbed by transverse circulation is not. Evaluate $\underline{A}_1$ and $\underline{A}_2$ and then the stress, neglecting terms of second order in u, v, and their derivatives. Some terms of the stress are of order zero (independent of u and v) and some are of first order (linear in u and v). Are the zero-order terms meaningful? What about the first-order terms?

7. Centripetal Effects.

Flow in circles is kinematically admissible in problems involving rotation of immersed bodies of revolution. In Newtonian flow these motions do not usually satisfy the equations of motion exactly, because of centrifugal force. The fluid is thrown outward where the centrifugal force is large, and circulates around and returns along the axis of rotation, where the centrifugal force is smaller. In viscoelastic flow, the fluid sometimes defies centrifugal force. For example, if a spherical body is rotated steadily, the fluid surrounding it may flow inward in the equatorial plane and outward at the poles.

Let us consider the circular flows $\underline{u} = r\omega(r,z)\underline{i}_\theta$, realizing in advance that such a flow is not likely to satisfy the momentum equation exactly. For these flows, $\underline{a} = \underline{i}_\theta$ and $\gamma\underline{b} = r\underline{\nabla}\omega$, so the stress would have the form

$$\underline{\sigma} = -p\underline{I} + r\eta(\underline{i}_\theta\underline{\nabla}\omega + \underline{\nabla}\omega\underline{i}_\theta) + r^2\underline{\nabla}\omega\cdot\underline{\nabla}\omega(N_1-N_2)\underline{i}_\theta\underline{i}_\theta - N_2r^2\underline{\nabla}\omega\,\underline{\nabla}\omega.$$

The term proportional to $\underline{i}_\theta \underline{i}_\theta$ is a tensile stress in the direction of flow. Since $\gamma^2 N_1$ is large in comparison to $\gamma^2 N_2$ and may even be large in comparison to the shearing stress, the circular streamlines act like they were stretched rubber bands. They tend to contract. Those with the highest tension (where the shear rate is largest) move inward. Those with lower tension may have to get out of the way. For example, with a rotating sphere, the shear rate is highest at the equator, so a stretched string of fluid around the equator will contract and accordingly move up toward the pole, if centrifugal force is not too large.

Let us add a fictitious body force $\underline{f}$ per unit volume to the momentum equation, whatever force would be required to keep the flow exactly circular. We can get some notion of the extent to which circular flow will be in error by seeing how big $\underline{f}$ has to be:

$$-\underline{i}_r \rho r \omega^2 = \underline{f} + \mathrm{Div}\, \underline{\sigma}.$$

With the stress as given previously, we obtain

$$-\underline{i}_r \rho r \omega^2 - \underline{f} + \underline{\nabla} p = \underline{i}_\theta \omega_r \eta + \underline{\nabla}\cdot(r\eta \underline{\nabla}\omega)\underline{i}_\theta - (\underline{i}_r/r)\gamma^2(N_1 - N_2)$$

$$- \underline{\nabla}\cdot(N_2 r^2 \underline{\nabla}\omega)\underline{\nabla}\omega - N_2 r^2 (\underline{\nabla}\omega\cdot\underline{\nabla})\underline{\nabla}\,\omega.$$

Supposing that $\underline{f}$ and $\underline{\nabla} p$ have no azimuthal component, the azimuthal component of the equation is

$$\underline{\nabla}\cdot(r^2 \eta \underline{\nabla}\omega) = 0.$$

(<u>Problem</u>: Verify this, as well as the preceding equation.)

Then the remainder is

$$-\underline{i}_r \rho r \omega^2 - \underline{f} + \underline{\nabla} p = -(\underline{i}_r/r)\gamma^2(N_1 - N_2) - \underline{\nabla}\cdot(N_2 r^2 \underline{\nabla}\omega)\underline{\nabla}\,\omega$$

$$- \tfrac{1}{2} N_2 \underline{\nabla}(\gamma^2) + (\underline{i}_r/r)\gamma^2 N_2.$$

Here the last term has been rearranged. (Verify.)

If we take N_2 proportional to η as an approximation, then since $\underline{\nabla}\cdot(r^2\eta\underline{\nabla}\omega) = 0$, we get $\underline{\nabla}\cdot(r^2 N_2\underline{\nabla}\omega) \simeq 0$ as well. There is probably not much harm in doing this since we expect the main effect of normal stress differences to come from the term involving N_1. Going back to the expression for the stress, we see that the terms in N_2 amount to an extra uniaxial compressive stress along $\underline{b}$-lines. Since $\underline{b}$ is roughly radial, this is not likely to produce much of a sideways force.

(Problem: Sketch curves of constant angular velocity for the flow in an infinite body of fluid surrounding a rotating sphere. Show the direction of $\underline{b}$ at various places.) As a final argument for neglecting this term, notice that we are nevertheless still retaining most terms involving N_2.

Noticing that the next-to-last term is a gradient, on rearrangement we obtain

$$-\underline{f} = [\rho r\omega^2 - (\gamma^2/r)(N_1-2N_2)]\underline{i}_r - \underline{\nabla}[p + \int_0^\gamma N_2\gamma d\gamma].$$

If the first bracketed expression on the right were a function of r only, it would be a gradient, so we could set $\underline{f}$ equal to zero and integrate to find p. In any event, for a qualitative discussion we can ignore the term involving p. The purely radial term is what prevents a solution with $\underline{f} = \underline{0}$. From it we see that centrifugal force and normal stress differences have opposite effects, assuming that N_1-2N_2 is positive, which is realistic.

Problem: Consider an infinitely long circular cylinder rotating steadily in an infinite body of fluid, with the fluid at infinity at rest. Show that all equations can be satisfied exactly. Next, put in a gravity force in the negative z-direction and let there be a free surface open to the atmosphere in the vicinity of the horizontal plane $z = 0$. Evaluate the stress component σ_{zz}, and find the locus on which it is equal to the atmospheric pressure. This will be roughly the shape of the free surface, provided that it is still fairly flat. Sketch the shape of the free surface. Explain the shape of the free surface qualitatively by considering that the circular streamlines are under tension and trying to contract.

Problem: In a cone-and-plate viscometer the fluid is contained between a flat plate $\theta = \pi/2$ (spherical coordinates r, θ, ϕ) and a cone $\theta = \pi/2 - \alpha$. If the angle α between cone and plate is very small, the shear rate is nearly the same everywhere. Assuming that this is true exactly, find the total normal thrust on the plate from $r = 0$ to $r = r_o$. If α is large there can be noticeable transverse flow. Which way would the transverse flow go if it were purely a result of centrifugal force? Which way if only streamline tension were involved? What is your guess about the actual flow pattern with both effects present?

8. Boundary Layers.

In classical fluid dynamics, flow at high Reynolds number past an immersed body is treated by supposing that viscous effects are important only in a thin boundary layer next to the body. The outer flow, away from the body, is treated as inviscid. In most cases the outer flow is a potential flow, i.e. $\underline{u} = \underline{\nabla}\phi$, with ϕ harmonic. Of course, there are viscous stresses in this outer flow region, equal to $2\eta_o\phi_{,ij}$ according to the Navier-Stokes equation. However, the divergence of the viscous stress disappears from the momentum equation because $\phi_{,ijj} = 0$. This is a great comfort, but it is not the real reason for forgetting about viscosity in the outer region. When we say that the Reynolds number is high, we mean that inertial effects are much more important than viscous effects. The stress is mainly the reaction pressure, which is whatever it has to be to move the rush of momentum around the body instead of through it. Even if viscous stresses did not disappear from the momentum equation, we would neglect them anyway.

For "slow" viscoelastic flow, in which γT is small, the Reynolds number may be large anyway. In the second-order approximation, potential flow still works. For, in potential flow, Div $\underline{A} = \underline{0}$ and thus Div $\underline{A} = \underline{\nabla}P$ with $P = 0$. Hence Div$(\underline{B}-\underline{A}^2)$ is irrotational, according to the result we obtained when talking about the second-order approximation. Moreover, Div $\underline{A}^2 = \underline{\nabla}\gamma^2$. Consequently, the second-order momentum equation is satisfied with $\underline{u}_2 = \underline{0}$ and

$$p_2 = -\frac{1}{2}\eta_o(T-2T^*)\gamma^2.$$

(Problem: Verify these statements.)

This is all very interesting, but just as in the case of the Navier-Stokes equation, it is misleading because we would still take the outer flow to be a potential flow even if such nice accidents did not happen; the stress is mainly the pressure $-p\underline{I}$ if the Reynolds number is large.

Within the boundary layer we can use the viscometric constitutive equation with good accuracy, at least if the flow is steady. For, the flow in a boundary layer is nearly a steady parallel flow. Moreover, the viscometric equation is exact, and not merely an approximation, at boundary points. We assume that the no-slip condition applies. In that case, the derivatives of velocity tangential to the boundary are zero, so if $\underline{n}$ is the unit normal vector, the velocity gradient has the form $\underline{\nabla}\,\underline{u} = \underline{n}(\underline{n}\cdot\underline{\nabla})\underline{u}$. This is of the viscometric form $\gamma\underline{b}\,\underline{a}$ with $\underline{b} = \underline{n}$ and $\gamma\underline{a} = (\underline{n}\cdot\underline{\nabla})\underline{u}$. We also assume that the motion is isochoric, in which case $(\underline{n}\cdot\underline{\nabla})\underline{u}$ has no normal component, so $\underline{b}\cdot\underline{a} = 0$ as required. The shear rate is constant in time for a particle stuck to the boundary because the particle stays at the same place and we have assumed that the motion is steady. Consequently, the stress is exactly

$$\underline{\sigma} = -p\underline{I} + \eta(\underline{n}\;\partial\underline{u}/\partial n + \partial\underline{u}/\partial n\;\underline{n})$$
$$+ (N_1 - N_2)(\partial\underline{u}/\partial n)(\partial\underline{u}/\partial n) - N_2|\partial\underline{u}/\partial n|^2\underline{n}\;\underline{n}.$$

Assuming that this is approximately correct throughout the boundary layer, we find that the stress component normal to the boundary layer is $-p - N_2\gamma^2$. Thus, under the usual assumptions of boundary-layer theory, $p + N_2\gamma^2$ is constant through the thickness, and equal to the outside pressure found from the potential flow solution. On the other hand, the "pressure" in the momentum equation (tangential component) is $-\underline{a}\cdot\underline{\sigma}\,\underline{a} = p - (N_1 - N_2)\gamma^2$. If p_o is the outside pressure, then the latter is $p_o - N_1\gamma^2$. This replaces p_o as the forcing term in the equation. Since the first normal stress difference is positive, the effective pressure is lower than if viscoelastic effects were absent. To see what the effect of this will be, consider the Blasius problem (flow along a semi-infinite flat plate). In this problem the outside pressure is zero (or constant). The boundary layer thickness grows like the square root of distance from the leading edge, $x^{1/2}$, so the shear rate decreases as $x^{-1/2}$. In the viscoelastic case, the effective driving pressure $-N_1\gamma^2$ increases as x increases. (It decreases in magnitude, but the sign is important). Thus the fluid is climbing an adverse pressure gradient. Whereas the Blasius solution for viscous flow shows no separation, in the viscoelastic case it appears that the fluid in the boundary layer might be brought to a complete stop at some point. From this example we may expect that separation will generally occur earlier when viscoelastic effects are important.

In spite of this, viscoelasticity may lower the drag on a body, because the drag is usually caused mostly by the wake. Elastic effects may cause the fluid to contract behind the body and squeeze down the size of the wake, thus reducing pressure drag. Such an effect would involve the response of the fluid to stretching motions, about which very little is known.

INDEX